SHARK STORIES

by
Al J. Venter
and Friends

Protea Book House
Pretoria
2012

BOOKS BY THE SAME AUTHOR INCLUDE:

Underwater Africa
Under the Indian Ocean
Report on Portugal's War in Guiné-Bissau
Africa at War
The Zambezi Salient
Underwater Seychelles
Coloured: A Profile of Two Million South Africans
Africa Today
South African Handbook for Divers
The Second South African Handbook for Divers
Challenge: South Africa in the African Revolutionary Context
Underwater Mauritius
Where to Dive: In Southern Africa and off the Indian Ocean Islands
War in Angola
The Chopper Boys: Helicopter Warfare in Africa
The Iraqi War Debrief: Why Saddam Hussein was Toppled
Iran's Nuclear Option
War Dog: Fighting Other People's Wars
Allah's Bomb: The Islamic Quest for Nuclear Weapons
Cops: Cheating Death: How One Man Saved the Lives of 3,000 Americans
How South Africa Built Six Atom Bombs
Dive South Africa/Duik Suid-Afrika
Barrel of a Gun: A War Correspondent's Misspent Moments in Combat
War Stories by Al J. Venter and Friends
Gunship Ace – The Wars of Neall Ellis, Helicopter Pilot and Mercenary
Guerrilla Wars in Southern Africa
Jihad in Africa

To Caroline,
my all-time favourite dive buddy and companion,
especially when there are sharks about ...

AND

To my dear and good friend Wolfgang Leander
who has made it his life's work to protect sharks.
'Wolfie' has ruffled a few feathers along the way,
but what the hell, old friend, it was worth it ...

Shark Stories by Al J. Venter and Friends
First edition, first impression in 2012 by Protea Book House

PO Box 35110, Menlo Park, 0102
1067 Burnett Street, Hatfield, Pretoria
8 Minni Street, Clydesdale, Pretoria
protea@intekom.co.za
www.proteaboekhuis.com

Editor: Danél Hanekom
Proofreader: Carmen Hansen-Kruger
Cover design: Bruce Gonneau
Front cover image: Zambezi shark on Aliwal Shoal – Photo by Roland Mauz
Back cover images by Paolo Fossati, Wolfgang Leander and Sijmon de Waal
Typography: Zapf Calligraphic, 10.5 pt by Bruce Gonneau
Printed and bound: Interpak Books, Pietermaritzburg

ISBN: 978-1-86919-692-9 (printed book)
ISBN: 978-1-86919-732-2 (e-book)

CONTENTS

FOREWORD

BY LESLEY ROCHAT

Also known as the Shark Warrior, Lesley Rochat is the founder and executive director of AfriOceans Conservation Alliance, an award-winning filmmaker, campaigner, activist, environmental writer, conservation photographer, and public speaker, who is fighting to save our sharks.

One day I met a shark named Maxine and almost overnight my life changed. No, it's not what you might think, I didn't lose a limb. Nor do I bear any scars, mentally or physically. In fact, I made friends with one of the most extraordinary animals on Planet Earth and also one of the most maligned.

Maxine is a raggedtooth shark, known in some parts of the world as the grey nurse shark. This one lived at Cape Town's Two Oceans Aquarium in South Africa. No ordinary shark, she has a remarkable true-life story that, for our purposes, started in 1995 when Maxine was caught in the shark nets of KwaZulu-Natal. Fortunately, she was found alive, promptly tagged and released by members of the KwaZulu-Natal Sharks Board.

Exactly 91 days later and 1,369 kilometres further down the coast, she was hooked during a fishing tournament. The anglers involved were aware that Cape Town's aquarium was looking for a shark of approximately her size, so they transported her to a tidal pool half an hour away on the back of a van. By the time she got there she was nearly dead, but fortunately she was revived. Maxine was later transported to the Two Oceans Aquarium where she was successfully introduced to the shark exhibits.

Six years later we had our first meeting. That occurred while I was photographing in the shark exhibit and the encounter would set in motion a course of events that would change both of our lives. I realised then that Maxine had the potential to play a pivotal role in shark conservation and, interestingly, that brief encounter resulted in my giving up my corporate career. I formed the AfriOceans Conservation Alliance (AOCA), a non-profit organisation, and developed the Maxine, Science, Education and Awareness Programme (M-Sea) for which this shark was the icon. In addition, I managed to influence a decision that ultimately resulted in Maxine's release into the wild.

We now had a 'Free Maxine', which can almost be likened to the Hollywood movie, *Free Willy*, except that ours is a true story.

Three years after my first encounter with Maxine and after nine years in captivity, this grand lady was satellite-tagged and in 2004, released back into the ocean. Her freedom made countless headlines, including international news, a television series called *A Shark's Odyssey*, a one-hour documentary titled *Maxine's Journey*, an untold number of radio shows, thousands of website links, educational books as well as public displays for children. She also inspired my multi-award-winning 'Rethink the Shark' campaign. In short, Maxine the shark was world famous – and that didn't happen because she bit anyone!

Maxine's release was followed by four more sharks being tagged and released as well as five wild sharks that were caught and tagged as part of an exciting research project that I was instrumental in designing. All these creatures joined Maxine in the sharks' Hall of Fame and became fantastic ambassadors for their species.

What is remarkable about all these events is that until then, sharks had been labelled 'monsters of the sea', possessed with insatiable appetites for humans. For that, this modern-day perception can be largely traced to the cult movie *Jaws*. In reality, that's all nonsense, especially since worldwide, only about five people are killed by sharks each year. In truth, irresponsible media tends to demonise sharks, and as a consequence, they are branded nature's outcasts. That, together with unconscionable and unsustainable fishing practices – in particular the barbaric practice of shark finning – has resulted in sharks being plundered on a massive scale in the oceans of the world. The statistics are staggering: it is estimated that over 70 million sharks are killed each year, principally by Asian longline fishing boats.

The plight of the planet's shark population cannot be viewed in isolation. The knock-on effect this slaughter is having on our oceans and on the environment is stupendous.

The implications of species lost and the largely unknown repercussions of what human greed, ignorance, abuse and neglect of our planet will cause, looms over all of us. Without being aware of our actions, we humans have become a new killer strain of cancer, slowly gnawing away at our host.

Scientific reports warn that conditions are going to get a lot worse unless we change our ways. If we don't, then the future of the human race looks bleak. Whether we can turn the tide of the environmental damage we relentlessly inflict, still remains to be seen. One thing is certain: we need to work together if there is to be any hope. Indeed, our survival hinges on it.

Each one of us must take responsibility for saving our planet's natural but limited resources. We need to start living consciously with the choices we make.

Our single greatest mistake has been to assume that we are superior to all other forms of life on the planet. In doing so, we have disconnected ourselves from nature: we have forgotten that by our very being we are a part of it and that everything is inextricably linked. Consequently, if our oceans are raped of sharks, which is inevitable at the current rate of slaughter, their disappearance will contribute towards our own collapse. We no longer have the privilege of time, nor do sharks. We *must* save these creatures so that we ourselves might survive.

Maxine captured my heart, and that, in turn, led me into what is now a comprehensive career in shark conservation. My initial objective remains the same today, to work towards raising awareness and to assist in their conservation.

When it comes to sharks, there is nothing more salient than the fact that we fear those things we do not understand, and also, we cannot care if we do not know.

My credo, consequently, is that I believe in educating people and effectively raising their awareness about the catastrophe that faces us. That, essentially, holds the key to changing attitudes, winning public support, and engendering political will, all of which are desperately needed to ensure their survival as a species: and in turn, that of all mankind.

The one message that this book holds, is a silent plea to all who read it, to ignite the ecowarrior within each one of us to fight and save our sharks.

AN INTRODUCTION TO DIVING WITH SHARKS BY AL J. VENTER

Diving with sharks has not always had the following it enjoys today. As American spearfisherman Carlos Eyles tells us in *Last of the Blue Water Hunters*, one of the outstanding books of the era and still very much perused almost 30 years after it first appeared, 'the shark, among all of the ocean's creatures, has been singularly responsible for keeping man's undersea activities confined to shallows of his shoreline.'

He goes on to tell us that our obsession with the shark's potential for killing 'is equal to our fascination with the spectre of death itself.' Yet he also makes the observation that it is only a tiny proportion of scuba divers that ever get to see sharks while in the ocean, never mind being threatened by these predators.

For all that, divers – and, for that matter, the public at large – 'are not immune to the books and movies or to the stories, real or fabricated, of attacks on humans by sharks. He has been conditioned, like everyone else, to react with terror at the sight of a shark, particularly if he is in the water with one.'

It's a process that needs to be undergone by anybody who contemplates encountering sharks in their natural environment, especially if they are going to be in areas where these creatures are commonplace, which means most tropical waters, and further afield as well. I recall reading of a New Zealand scuba diver being attacked and killed by a great white shark in the waters around Antarctica. I also used a photo in one of my earlier books of an immense great white being snagged in nets off Malta. Who would have thought?

Eyles says that process began for him when he had 'an encounter with the devil himself'.

He'd been diving blue water for almost two years and was filled to the brim with all those culturally induced fears. As he explains, while one eye looked for fish, the other attempted to pierce the impenetrable blue looking for what he likes to refer to as 'eaters', as the underwater tribe likes to call them.

'I had speared a yellowtail off Pukey Point at the Coronado

islands, out of San Diego in Mexican waters, and as I worked the fish to me, a ten-foot dusky, one of the few aggressive sharks in these waters, stormed in and hit the yellowtail just behind the pectoral fin, violently tearing off the body, and swam away. The speed and ferociousness of the assault was everything the films depicted, and I unwittingly transferred the scene onto myself. The dusky graphically demonstrated what I had long felt, that man is woefully vulnerable in the ocean.

'That vulnerability and the realisation of it are among the primary reasons why there are so few blue water hunters. As the race car driver must accept the hard reality of an accident in a race, the blue water hunter must accept his own vulnerability in the ocean.

'But there the similarity ends because a racer does not have time to think about an accident; he is fully occupied with driving the car. The hunter has plenty of time to think. He thinks about the speared fish and how the bloody and frenzied movement calls out for whatever lies lurking beyond his range of vision. He ponders his slow movements while travelling from the surface to the depths and back, knowing he moves like a sick or wounded seal. He is acutely aware that he would be unable to flee or hide from an attacking shark in the improbable event were he given the chance. He carries into this alien environment nothing more than a bow and arrow with which he must protect himself against a prehistoric beast whose overpowering size, strength and manoeuvrability only further accentuate his own glaring inadequacies in the ocean.

'The dusky swam in my thoughts for a long time. I couldn't bury the scene of it tearing into the yellowtail. This preoccupation eroded my concentration and stole the joy that comes with blue water hunting. I hoped that in time the shark would fade and I'd feel as free in the deep water as I felt in the kelp beds. But it did not leave me and I felt its weight whenever I moved into bottomless water. Over the next year I carried that dusky until the excess weight of it started to drown me. The only escape from a situation that was getting worse instead of better was a clear or final decision: either I must release the shark completely or quit diving and do something else.'

He does admit that the prospect of giving up diving because of that single encounter was unthinkable.

'I had to pry myself loose from the shackles of the dusky. The first step was to accept my vulnerability in the ocean, that is, clearly accept the reality of it without the influence of fear that muddies the water and feeds the terror. In that full acceptance lies a certain freedom; a new space is created where there is room to manoeuvre or defend, or even attack, if necessary, much the same as a full acceptance of death alters the view of life and brings a certain lustre and richness to it that was not there before.

'Secondly, I had to erase from my mind all images of sharks that my imagination conjured. That was the key; when the sharks began to filter into my head, I'd block them out by focusing on the ocean before me. It was not easy at first, but it soon became habit, and eventually my mind cleared itself of its dark wanderings, and my freedom was nearly complete.

'As thoughts of the shark faded, the ocean opened up to a new and broader view. My concentration improved and my hunting skills followed; soon my kinship with the blue water became equal to that of the kelp beds.

'It was during my second trip to Guardian Angel Island that I had just speared a large grouper. The fish had had its way and had holed up in 60 feet of water. Diving to the cave, I hung at the entrance trying to determine the best way to pull the grouper out. Suddenly and very clearly I felt another presence in the water, and turned in response. Behind me, 20 feet away, was the largest shark I'd ever seen before or since in the water. Face to face it was huge, a yard wide in the head and eighteen feet in length. It is difficult to judge the weight of a creature of this size; a guess would be somewhere between 1,500 and 2,000 pounds. It hung motionless in the water with all the confidence of an old despot.

'While the shark's great power was oozing through its skin, my speargun stock was floating on the surface. Out of breath, I ascended; it was a vulnerable moment. The shark made no move towards me, and when I reached the top, it swam lazily off and out of sight.

'After resting for five minutes, I dropped back down to the holed up grouper. The fish had worked its way deeper into the cave and as I tugged away on it that same feeling came over me again, clear and clean as being gently but firmly informed that something was about. Turning, I found the same shark ten feet behind me. Dropping the line to the fish I made a half threatening pass, empty-handed. It was a weak gesture and the shark did not move an inch. Needing air, I pushed off the bottom and began to kick my way up. As I did, the shark moved into me. Kicking harder, I rose above it and as I did so it swam directly beneath me, so close that I had to lift my fins to keep from hitting its massive dorsal fin. When I reached the surface, the shark turned and came back toward me. Considering its size, I couldn't interpret this move as anything less than aggressive.

'Spotting the anchored boat a hundred yards away, I swam to it, keeping an eye on the shark as best I could until it broke contact with me. Fifteen minutes later four of us returned and buzzed the area in the boat. Then several hunters rode shotgun for me while I retrieved the grouper. The shark made one last fleeting appearance and then was gone.'

Which is a marvellous end to Carlos Eyles's wonderful real-life adventure.

When you've dived as long as I have, I suppose you'd like to think that you get to know something about sharks and shark attacks. Purposefully – and sometimes inadvertently – I must have encountered sharks underwater scores of times in a diving career that spans more than 40 years, most times either on or in the vicinity of Aliwal Shoal or in Cape waters.

I've had four friends who have been actual victims of shark attacks, five if you include Conway Plough's buddy with whom he was doing an underwater repair job in Mombasa's yacht harbour at the time. Conway had his leg seriously ripped by a bull shark – *Carcharhinus leucas* and commonly referred to as the zambezi shark in Africa's 'Deep South' – and though it required years of treatment and rehabilitation, he went on to run marathons, which shows what personal application and determination can do.

Others who were attacked and survived include Tommy Botha who had a great white come up at him from below – as these creatures quite often do – and usually at great speed. The beast hurled him clean out of the water. His wounds needed multiple stitches, though he'll tell you today that the critter was only 'mouthing' him. Had the white gone in for a kill, Tommy reckons it was easily big enough to have ripped him in half had it a mind to do so.

The same with my old dive buddy Willie van Rensburg of Hermanus. After years of poaching crayfish – and even diving for these crustaceans off Dyer Island – he ended up with parts of his anatomy inside a great white while sport diving with his sons. I deal with that little episode in Chapter Nineteen.

What is notable about that event is that a few years before that happened, I'd been along on a rather delightful shark-hunting expedition organised by two old friends, Jeff McKay and the American marine artist Richard Ellis who specialises in maritime artwork, mainly for Washington's *National Geographic*.

Jeff has since hung up his flippers in favour of flying commercially in Australia, but he remains extremely knowledgeable about all species of shark, having worked years for the KwaZulu-Natal Sharks Board. Richard Ellis's book, *Great White Shark*, was first published by the Stanford University Press in the United States in 1995 and produced in conjunction with John McCaskey. It remains one of the classics of the genre.

We'd hired Traill Witthuhn to provide us with his boat, to which we attached a shark cage, the only time I was ever to attempt to dive sharks from the safety of a steel cage.

Though we had the boat and the requisite cage, and spent a week at sea, absolutely nothing happened. While we made one of the local butchers happy by buying a couple hundred litres of ox blood and offal, all of which we dutifully dumped into the sea at regular intervals, we never saw a single shark, never mind a lonesome great white. And that in waters off Dyer island that these days attract thousands of tourists to view these denizens from up close.

Geoff joked afterwards that perhaps one of us should have fallen into the sea. That would almost certainly have attracted sharks, he quipped.

We never did test that premise.

Another shark-related incident, not yet in the record books, involved Walter Bernardis, with whom I spent weeks along the KwaZulu-Natal South Coast. In the winter of 2006, we dived among packs of blacktips and other critters that had come in for the chum that we had suspended in the water to lure them up close.

There would sometimes be dozens of them, and while the situation deteriorated into a kind of feeding frenzy, we dived among them and hoped that we weren't attracting too much attention. Sharks were everywhere, above, behind, ahead and below.

I remember very clearly Walter telling us to keep our hands close to our bodies as these predators weren't all that discriminating. They would go for anything that protruded – loose air hoses included. A hand out of place would have been especially tempting, which is why photos of those dives always show us with our arms folded ...

I deal with some of those little episodes in Chapter Eight, and also tell of an incident that came pretty close to changing Walter's lifestyle when a four-metre tiger shark tried to engulf his leg. He'd committed the one 'crime' that he always warned us divers against.

'Remember,' Bernardis would say. 'The shark is a predator. It interprets the actions of those it encounters underwater in exactly the same way that a predator like the lion and the leopard would do on land. So while you might have sharks behind you while you're suspended in the water, you never turn your back on a shark and swim away.' That would signify fear and retreat, he stressed. It was also when predators go in for the kill.

Walter was neither fearful nor was he retreating from the monster that very briefly latched onto his leg, causing an incident that also required a load of stitches. He'd simply finished what he was doing and was headed back towards the boat.

In different circumstances we also had with us British national and Durban resident Len Jones who had a number of encounters with sharks. That was at a time when spearos would string the fish they'd shot onto their belts. What was astonishing about this was that over a period of years, Len Jones and his buddies would keep up that same routine and they'd very rarely have problems.

Len was only bitten once and it could have been much worse. It was a milestone event. It also signalled to all South African spearos from then on that if you shoot a fish, you don't hang it from your belt. The drill these days is to have a float attached to the end of a line – often to an inner tube – and you haul that lot behind you while on the hunt until you eventually leave the water.

There are a lot of other friends – spearos, mainly, who have had run-ins with sharks. I don't know of one who hasn't experienced sharks that were more curious or aggressive than others at some time or another. Darrell Hattingh has had dozens of encounters. So has the rest of the spearfishing gang, but these are all good divers, many of them having made national – and in some cases, international – spearfishing championships.

While my own experiences with sharks might sound impressive, they are not. I have friends who have dived with these creatures on Aliwal Shoal and Protea Banks ten times as often as I have. Though there have been a few tense moments, attacks have always been rare – extremely so – but we've all been 'buzzed' at one time or another.

The worst offender here is the bull or zambezi shark, which likes to get right in your face if it senses any kind of uncertainty. This is why I *never* dive in murky water, especially when you have rivers emptying out their dirty effluent into the sea after the rains. That's the kind of environment that zambezi sharks like, though even then they sometimes ignore obvious prey.

We had one aspiring spearo drowned off the Umkomaas River

mouth not very long ago. The seas had been rolling in big time and being an upcountry visitor, he couldn't tell good seas from bad. In fact, he shouldn't even have been in the water it was so rough. So it was not all that surprising that he disappeared, drowned presumably.

His body washed ashore two weeks later, badly bloated and clearly got at by crabs and other creatures that inhabit shallow waters. But there were no big tears on his body, no evidence of shark attack.

And that in semi-tropical waters that have the more than usual quota of zambezi sharks ...

In fact, my worst shark experience involved a zambezi shark. I'd arrived back in South Africa from my home in the United States. The idea was to gather more material for the book on which I'd been working just about forever.

For this purpose, I'd arranged a shark dive on Protea Banks with my old friend Graham Powell, who, with his wife, ran one of the best-known dive facilities in Umkomaas on the KwaZulu-Natal South Coast. Trouble was, I'd had a steak on arrival in Johannesburg and within hours, was seriously laid low. The doctor said afterwards that it was food poisoning.

I'd entered the country on a Saturday and the trip to Protea Banks was scheduled for the following Wednesday. Meantime, I battled through what must have been one of the worst salmonella attacks ever, which is saying a lot for somebody who routinely travels across a range of Third World countries and, as a consequence, over a lifetime of activity in these places has acquired a powerful antibody defensive system.

I was still not well when Graham took me on an orientation dive on the wreck of the *Produce* on Tuesday. It was a great dive but I went straight to bed afterwards. We left for Shelly Beach, about an hour down the coast, very early the following morning.

Protea Banks, about eight kilometres offshore, has always been rated as one of the world's best shark-diving locations. Though things have changed, it was only the other day that you'd go down and encounter a dozen or more zambezi sharks during the half-hour that you drifted along the gradually up-sloping reef. Go into the water there today, and though there aren't that many 'zambies' around, there are still a lot of sharks. Caroline, my shark-loving partner, dived on Protea Banks in April 2012, and on her first dive off Shelley Beach she almost landed on top of a pack of 14 large guitar sharks. Interestingly, the last time I saw guitar sharks was 35 years ago on Aliwal Shoal.

There was always a lot of marine life, which makes the banks one of the most popular game-fishing venues in the country. Its 'residents' include spotted eagle and manta rays, several species of hammerhead, the occasional tiger shark and sometimes more raggies than one would find on a normal day on Aliwal. Add to that tally raggedtooth sharks, threshers, duskies, coppers as well as a variety of sand sharks. Now and again a diver will return to the surface and report a sighting of a great white. In Protea Banks I discovered a treasure that was to be cherished.

Because of its sharks, the banks were always considered to be something of an advanced dive. The water was deep, there were sometimes stiff currents and that meant that divers needed the right kind of qualifications. Generally, the accepted norm was somewhere near instructor level. Eventually, tour groups would end up taking people who could prove that they had a minimum of 20 dives.

The day I got there with Graham and five others was no different. The routine was that we should all exit the boat together and instead of gathering on the surface and giving each other the usual OKs, we'd head straight down the 35 or more metres to the bottom instead. Once below, and as the sharks started coming in, we'd do the checks. With Caroline's recent dive, divemaster Roland Mauz insisted on the group going in on negative entry.

For me, things went bad from the start. I'd been lax in checking my weights, which meant that I was slightly buoyant. That didn't matter at any kind of depth, but it made a difference as we headed towards ten metres. Also, I was still feeling bad.

The ten-minute boat ride out didn't help either. It shouldn't have surprised me that the first group of sharks to take an interest in our group came right in and at speed. It was the usual thing, head on and veering off at the last moment, sometimes three or four in a row.

Looking back afterwards, I would recall that the predators of the world – sharks, the big cats of Africa and Asia, and many other creatures besides – are physiologically geared for exactly that kind of condition. Call it instinct or perhaps the feral nature of the beast, it's an essential aspect of the hunt. It is also what makes them tick: why they target one animal in a herd in preference to others. Still not fully understood by science, the world's predators have their own built-in detection systems that isolate the sick, the lame and the aged from their strong and healthy counterparts.

Suffering from the after-effects of food poisoning, I fell very much in the former category. That was unfortunate for both me and for Graham. As my dive leader, it was his job – like it or not – to offer me backup.

Things never really let up during all the time I was under. We'd have a few minutes of respite and then the sharks would come in again, though I was never actually physically harmed. It was disconcerting, nevertheless – what some would call a close shave ...

Eventually, with the boat closely following our bubbles and our progress underwater, Graham signalled once our group had reached shallower water, that I should leave the group and return to the surface. In retrospect, that was the longest and loneliest 20 metres I've ever swum on my own. Curiously – and totally contrary to what I was expecting – once away from the main group, none of the sharks followed. Most disconcerting of all was the time the guy on the boat took to hauling my gear on board before he gave me a hand.

An interesting sidelight here is something that happened while diving on off Aliwal Shoal and involved tiger sharks. One of the licensed shark dive operators had taken a group down to the bottom to view a pack of these creatures that had been attracted by chum, when a modest-sized tiger shark, apparently new to the area because it had never been spotted before, swam down and bit one of the divers in the head.

Had it been the kind of attack that tiger sharks are better known for, our unlucky friend might have been decapitated. Instead, it simply 'mouthed' the victim's head – or, in the lingo 'tasted' what was on offer. Then it swam off again. It says a lot that the diver wasn't badly hurt.

The 'testing' or tasting of a potential prey by sharks – or, as it is professionally termed, 'mouthing' – is a feature of the kind of feeding routines that are practised by some sharks, the bigger ones especially. Essentially, mouthing a potential victim establishes whether the individual or object might be edible. Other sharks quite often brush against the intended victim, their skin sensors telling them whether the object or person would make for a meal, which is why so many swimmers, just prior to having been attacked, felt something underwater brush against them.

With 'mouthings', the shark evolutionary process long ago indicated to these creatures that it is preferable to 'taste' something that looks appetising before delivering the coup de grâce. That might involve a bite of several hundred kilos per square centimetre.

It is also why Tommy Botha wasn't mutilated and possibly killed when he was mouthed by a great white while spearfishing in Cape waters, something I deal with in Chapter Twenty One: sharks apparently don't go for wetsuit neoprene.

The impact of the shark coming at Tommy from below, almost like a missile, caused a lot of damage. His wounds needed scores of stitches. But he was alive. In fact, he was able to swim to safety, and it wasn't long before he was spearfishing in those same waters again.

Had the white snapped the full force of its jaws, this former Springbok spearfisherman would almost certainly have been listed as a statistic.

Like great whites, zambezi sharks seem always to have been in the news. Graham Lambert has some excellent stories about real events involving zambies on his website. The majority of postings are obviously limited to survivors' tales, though some of the events listed on *The Shark Files* make for compelling viewing.

One attack took place in 2004 and involved another KwaZulu spearo, Warren Bennet, who'd gone diving on a bright summer afternoon after work and found himself in about 20 metres of water. Having tried to get in close to some yellowbellies, he spotted a 'cuda of about 18 kilograms, followed it down and shot it.

'The fish took off with great speed, pulling me around for five minutes. When the fish finally gave up the ghost, I swam down to retrieve it ... which was when a large zambezi of over 150 kilograms raced in, directly in front of me. It showed no regard for my presence.' Having sunk its jaws into the fish, the shark ripped Warren's spear from his grasp and dragged spear, gun, fish and float underwater for about 20 metres before the spear pulled free from the crippled 'cuda.

The shark, now beyond the spearo's immediate visibility, led Warren to believe that it had swum off happy with its easily-acquired dinner. The next moment it was right there again, almost in his face.

'It buzzed me for a minute or so, swam around me, after which it settled about eight metres below, just circling. By then it was starting to get dark so I decided to call it a day. 'The shark followed Warren for about 400 metres all the way into shallow water. It finally turned around and headed back out to sea.

Looking at the list of shark stories that have appeared in print, both in book form and on the web, one could conclude that attacks appear to be divided almost equally between zambezi sharks – or bull sharks, as they're referred to abroad – and the much more newsworthy great whites.

Speaking empirically, especially since I've dived great swathes of ocean along the east coast of Africa — almost from the Somali border southwards, as well as the Red Sea — my guess is that there are many more attacks by bull or zambezi than this species of predator is generally given credit for.

There was a time in the 1980s when nobody would dare venture into the waters around Mogadishu. Shark attacks were taking place weekly, though very few ever made the news because Somalia is that sort of place.

I covered the 1993 American-led United Nations Operation in Somalia, or UNOSOM as it was officially termed. There were several thousand American military personnel in the country at the time, together with a string of US Navy ships lying just offshore. It astonished me all the more when I spotted a dozen or so US Marines heading towards the lagoon on my first afternoon there, having earlier in the day been flown across from Mombasa in a USAF C-130.

When I mentioned to my military escort that there had been numerous shark attacks in that area just a while before, he smiled condescendingly and changed the subject. He obviously didn't believe me. For all my protestations, American soldiers – male and female – continued to swim in those waters until Somalia was finally abandoned after the incident which led to events recounted in Mark Bowden's excellent book *Black Hawk Down*.

Nor was this an isolated incident. At the time, Kenya was being plagued by shark attacks, though very little of it ever appeared in the press. Tourists were the mainstay of Kenya's economy and simply too important a financial issue to draw attention to something like man-eaters.

Then something happened that I remember as if it were yesterday. I haven't had many open-water experiences with great whites, but the few times I did run into these critters were always memorable. That dive on the *Maori* was no exception, because it was one of two dives where I had 'contact' with a great white shark.

It was a family day for us: my wife and children together with a small group of friends headed out from Hout Bay – a fishing port and suburb of Cape Town – to the wreck and what promised to be a crayfish fest (rock lobster in the northern hemisphere). The only difference was that we had to catch the crustaceans first.

While the sea was calm and cold, it was winter and a storm the previous day seemed to have churned up the water. That meant that we had a visibility of about two metres, at best. Normally, when out with strangers and considering the depth, we might have aborted, but not this time. We'd all dived the old lady before and having progressed that far, there was no turning back. Anyway, the skipper was insistent about his *kreef* ...

Five of us entered the water and descended rapidly. Once on the ship we split up, me poking deep for something to take back to the surface. In the end I'd snagged half a dozen crayfish and was the last of the bunch to head for home.

I must have surfaced about three or four metres from the boat, where I could see that the cooking pot for the 'bugs' we'd collected was already bubbling on deck. On coming alongside, I was faced with a long metal pole to which had been welded strips of steel to form a makeshift ladder that hung over the side. The only difference was that instead of something conventional, the 'steps' stuck out on each side of a heavy metal rod.

That prompted me into doing something that I'd never done before while still in full diving kit. Instead of removing my tank and weight belt, as I'd customarily like to do, I decided to go straight up the ladder with all my gear still on, one step at a time. By then, my family – Madelon, as well as Luke and Leigh — were watching my clumsy progress with interest. Finally, I was able to swing a leg over the transom.

At that moment, barely an arm's length from where I was perched and not yet quite on board, a huge great white shark surfaced alongside where I had been moments before. It stuck its massive head out of the water – it was almost a metre across — and with one of its great big black eyes, gave me the once-over for a few seconds.

Then, as unobtrusively as it had arrived, the monster slipped back into the sea and disappeared. Somebody cracked a joke about one of us perhaps going back into the water to get more crayfish.

We agreed afterwards that the predator had obviously been in the water all the time, probably throughout our dive. Had any of us spotted it while still below, the visibility would have been too poor for us to have seen its length. As it was, with such a large head, it must have been something like five metres long.

Looking back, and in the light of some of the great white attacks on divers in False Bay since then, I suppose this big mother could have attacked any of us at any time. But it didn't. Which reinforced the views of some of the Cape professionals with whom I've dived in the past, people like Jerry Buirski, Tommy Botha and Mike Rutzen. All maintain that great white sharks are pretty fussy what they sink their jaws into.

Heading back to Hout Bay after lunch, it was Luke who later pointed at what was left of a seal that had been bitten in two. The upper half of its torso floated in quiet water about a kilometre from where we had been moored. It had probably been killed a short while before because there was still a splash of fresh blood in the water around the remains of the carcass.

As Madelon commented at the time, it might even have been the same shark ...

There have been other incidents. With scuba equipment, I once dived in an area to the immediate south of Fish Hoek, an area not routinely frequented by the average underwater enthusiast. It was perhaps because of the isolation that we decided to give it a bash: who knows what lay off the rocks opposite Glencairn railway station.

We'd been finning along for about ten minutes, headed in the direction of Fish Hoek in a none-too-inspiring underwater environment, when suddenly, from behind, a huge great white swam at an incredible speed between the two of us. There had been no warning and certainly no indication that it was interested in either of us. In fact, my immediate impression was that the shark's attention was directed at something ahead.

Though the water was relatively clean, it disappeared into the murk before we were able to establish what it was after. Not that we hung around very long after that happened. Coming from behind, it could easily have made a meal of both of us and we would have been none the wiser before it was too late.

Judging by the number of subsequent great white sightings in the Fish Hoek area – coupled to several fatal attacks on holidaymakers – the area seems always to have had its resident shark population. Which makes one speculate that since there are swimmers in the water just about every single day of the year, what was it that triggered the latest shark activity?

Rob Erasmus, a NAUI diving instructor with 20 years' diving experience, doesn't provide an answer to any of these questions, though his own experience in the area is instructive. As an aquarium volunteer diver who has a keen interest in sharks, nudibranchs and underwater photography, he'd offered his services to the South African Museum to collect specimens for their catalogue collection. One October they asked if he could supply them with some hermit crab material.

Having dived in the vicinity of Fish Hoek often enough before and familiar with conditions, Rob believed it would be a bit of a breeze. It wasn't. As he recalls, diving conditions off Sunny Cove were perfect, the water being 16 degrees Celcius and the visibility in excess of ten metres. The dive was planned to a maximum of eight metres for 40 minutes, never further from the shore than about 30 metres.

He takes up his story: 'Returning along the bottom after about half an hour, something made me turn around. I was greeted by the vision of a very large shark. My mind went into an automatic shark mode ... it was swimming directly towards me.

'Deflating [my] BC and exhaling so that I sank to the bottom was my only option, after which the shark passed about a metre above. It was a very large female that I estimated to be somewhere between four and four and a half metres. I also noticed that her left pectoral fin had a large portion of its tip missing.

'She circled me and lined up for another pass. Believing she was showing more than general interest, I reckoned that it warranted some form of protection, which was when I quickly backed over the flat rocky bottom towards a small crevice.

'The problem, however, was that the crack was only large enough to protect the upper part of my body. By now the shark was heading in for a second pass and was a bit lower than before.

'Once again the shark passed overhead, *less* than a metre away this time and there was no doubt, she really meant business. With my mind racing, I knew that if she came in any lower, there would be no chance of my escaping her. I was also worried that I might do something to antagonise her, like gesturing, waving my arms, blowing bubbles or possibly screaming. I knew I had to remain as calm as possible and was aware that if she did decide to attack, there was nothing that I could do. That's how big she was.

'The great white circled again and came towards me on her third pass, this time directly at head height. She was about three metres away when she opened her mouth and pumped her gills: I was looking directly into her massive jaws. At the very last moment, with the shark about a foot or so above me, I put out my hand and ran it all the way down its belly.'

As Rob tells it, it was the last time the shark came for him. She simply disappeared into the blue. Following another five minutes' wait, he broke cover and after a hasty swim along the rocky bottom, reached the shore.

Having contacted numerous shark experts and related his Fish Hoek experience, several confirmed that a large female white with the left tip of the pectoral fin missing had been spotted off Seal Island, False Bay, some months previously. The consensus was that the shark was a bit smaller, probably just less than four metres.

Rob Erasmus seems always to have had an interest in sharks and has been diving since 1984. He has been on a number of environmental courses registered with NAUI and his experiences range from underwater tagging of raggedtooth sharks in the South Cape area to cage diving among whites in Mossel Bay.

'What I did learn is that great white sharks occur close inshore in False Bay, in this case less than 30 metres from the shore, and in very shallow waters,' he told me and went on: 'This one was particularly bold. She gave me the impression that she'd seen and done it all before, which meant she was very much in control of the situation.

'I can only conclude that my touching her probably caused the creature to move on. My advice to divers when faced with something like this is to stay alert and calm.

'To antagonise any shark that size could easily result in a physical encounter that might end up in disaster.'

Wolfgang Leander

CHAPTER ONE

THE MOST INCREDIBLE SHOAL ON EARTH: THE SARDINE RUN

Debbie Smith, for years one of the stalwarts linked to diving in the Seychelles, returned to South Africa to found a specialist ecotourism company called 'Diving with Sharks'. During this time she has been involved with what the pundits like to call the 'sardine run' and she has so done more than most. As she says: 'The run is quite the most remarkable experience in any diver's life.'

Durban-based Peter Pinnock, one of the world's most notable underwater lensmen, agrees with Debbie Smith. He admits that some of his most unusual diving moments – in a career that spans decades – came from diving off the South African coast during the annual migration of sardines.

He recalls a dive that took place off the Wild Coast recently. 'The sea around the boat had turned from blue to dark brown as an enormous cluster of sardines, perhaps 80 metres across, passed under us ... the water seemed to boil as these tiny fish were furiously chased to the surface by predators – sharks, porpoises, barracuda and the rest. The sardines ducked this way and that, in a kind of unrehearsed choreography that only nature could provide.

'One moment these clusters would swarm on the surface, the next they would retreat, as in a single living mass, to deeper waters. When the big fish gave way, cormorants, gannets and other birds went into the attack. Like the predators, they were relentless, diving into the water from great heights, feasting and then taking off again for another go.

'Having lingered briefly on the surface, the birds left behind a rather distinct odour ... and through it all, there were sardines everywhere.'

The annual migration of the sardine (*Sardinops sagax*) from the Agulhas Bank of the South Cape, past the Wild Coast — in what was once known as the Transkei — and up the coastline to KwaZulu-Natal, takes place during the southern hemisphere's winter months. Nobody knows where these little creatures start their lengthy, suicidal journey, or for that matter, where it ends, except that it is a natural phenomenon and takes place almost every year. In 2003, the sardines failed to 'run' for only the third time in a quarter century.

This most unusual spectacle – the only one of its kind on the planet – is still poorly understood from an ecological point of view. There have been various hypotheses – some of them contradictory – that try to explain why and how it all happens. Sardines, or as they are sometimes called, pilchards, are several types of small, oily fish that are related to the herring family *Clupeidae*. The accepted wisdom is that those under six inches (plus or minus 15 centimetres) are sardines: anything larger (and of the same or similar species) is a pilchard. But then, nobody makes too fine a point of it.

According to Debbie Smith, a recent interpretation is that the migration – for that is what it amounts to – is most likely a seasonal reproductive movement of a genetically distinct subpopulation of sardine that moves along the coast. Unusual in both its extent and duration – it can last months – this natural marvel is rated to be the marine equivalent of the famous East African wildebeest migration that takes place on the Masai Mara and Serengeti Plains each year. Those few who have seen both, maintain that the sardine run is just as thrilling and spectacular to witness.

Though it has been with us almost forever, it is only in recent years that the diving community realised that the event was unique enough to attract underwater enthusiasts from all over. Indeed, underwater they have arrived in South Africa from all over the world, paying minor ransoms to be allowed to experience the sensation not only of diving among the bait balls, but also observing other marine creatures that congregate at migration time, sharks especially.

The sharks arrive in their thousands, as do other sea creatures, and curiously they take little notice of small clusters of divers around the fringes – and sometimes inside – the bait balls. In fact, while the business side of the sardine run has been an annual feature, there has never been a fatal attack, though most of those involved with the spectacle maintain that it could still happen. While there are millions of sardines in the water, says Debbie Smith, 'the sharks are quickly sated: they don't have to view members of the diving community as potential meals.'

There have been several incidents of course, most of which stem from a bait ball the size of a building suddenly changing direction in the water, usually because of the presence of predators.

Divers in the water along the fringe of such a bait ball will suddenly be enveloped by this moving, stifling accumulation of small fish. And while they are totally surrounded by this mass, they will be effectively 'blind' to any predators charging through with their jaws agape. It is the kind of thing that happens all the time, and has resulted in several shark bites or tears, none of them severe enough to cause loss of limbs, but often serious enough to call for medical treatment.

In one case during the run in 2010, Walter Bernardis had a German client who made a point of trying to become enmeshed in bait balls. Though he was repeatedly warned of the dangers by Walter, the man – clearly trying to impress his mates – persisted until he was bitten in the face by a passing shark. It was something that happened quite by accident, because at the time he was in the middle of a huge shoal of sardines and the shark was probably as 'blind' as he was.

'We had to work pretty fast with this guy once we got him onto the boat. As with most head wounds, there was blood everywhere and he was bleeding badly, his nose still barely attached to his face,' Walter recalls. The 'victim' was taken ashore to a local clinic where his nose was attached again and his other wounds stitched up with

Paolo Fossati

3

what appeared to Walter to be some kind of extra-strength fishing gut. 'It was clear that the Third World medical professional had very little experience of this kind of injury,' he recollects.

'So one of my men who was heading back to Durban anyway, rushed him to hospital where a bunch of doctors, as well as a plastic surgeon, sewed his nose on again, properly this time.'

Walter has since heard from the man, who even sent him photos of his face from Germany a year after this dreadful event, which, he acknowledges, could have gone seriously wrong. 'The guy is fine, at least judging from the photos we received ... you can't even see that his face had been ravaged ... bit of luck that ...'

Most diving activity during the month or two that the season lasts, centres on the small time-forgotten coastal village of Port St Johns, situated on Pondoland's magical Wild Coast. What the town lacks in modern comforts – though there are good restaurants and hotels – it makes up for in location. The region is steeped in cultural traditions that allow the Amapondo people to live in healthy isolation from other Eastern Cape tribes and protect their cultural heritage.

Peter Pinnock tells about his own search for sardines and mentions driving to the Mkambati Nature Reserve in the Eastern Cape in the hopes of a first sighting as they approached the coastline.

'Getting to Mkambati, a former leper colony, was an adventure,' he tells us. 'The two-hour drive from Flagstaff along a twisting, bumpy dirt road left our kidneys aching and somewhere along the way one of the car's bumpers fell off, never to be seen again. Having reached the village of Mkambati, we joined a distinctly dejected-looking bunch of film-makers and nature lovers who for three days had been carefully monitoring the open sea; until then, not a single shoal had been spotted. Somewhat disillusioned at the lack of action, we joined their ranks.

'At first light the next morning one of the pilots, Paul, took off in his microlight, headed south down the coast and flew in the direction of Port St Johns. Along the way he spotted a few small isolated pockets of the fish, but nothing substantial. When this information came in, we decided to head north towards Port Edward. Paul said that we should use our boats, which we did.

'After securing our equipment on board, it was an easy launch through the Mzikaba River estuary and onto an ocean as placid as glass. The skipper set a course along a coastline that remains one of Africa's undiscovered paradises and along the way, we were able to view areas that are mostly impossible to reach by road. Much of the shoreline was pristine, primeval almost. *Spectacular* would be an appropriate description.

'There was hardly anybody to be seen along myriads of coves, cliffs, chasms, inlets and beaches, most of which are almost as untouched and unfrequented today as they were when the Portuguese navigators, headed for India five centuries ago, came upon them.

'In several places we saw eland – Africa's largest antelope, which the local natives sometimes refer to as "mountain of meat". These graceful creatures grazed on cliffs that plunged down hundreds of metres into the sea. At Horseshoe Falls, a river cascaded into the ocean below: a ribbon of water that had crossed a corner of Africa originally made famous in the book *Cry, the Beloved Country* that brought fame and acclaim to Alan Paton.

The main area of sardine activity along this stretch of coast during the season runs from Port St Johns to Mbotyi, with the oceanic continental shelf running close to land. That allows deep water to run close to shore, and, in turn, concentrating both the cold current that wells up from the south during the winter months and warmer waters coming in from the north down and the Mozambique Channel and pushing them closer to shore, condensing the shoals.

As these immense shoals move up the coastline, they feed in nutrient-rich waters with literally thousands of larger fish in attendance, feeding off them. It is a fairly vicious cycle, with the sharks and other predators constantly harassing and splintering pockets from the main shoals and devouring them at their leisure. This is one of the largest congregations of ocean predators known to man.

The single most impressive feature of the run, unquestionably, is what the pundits like to refer to as the 'bait ball', a living, moving, pulsating, and energy-generating conglomeration of marine life that can sometimes be as wide as football pitch. Anybody involved with the sardine run, will constantly hear the term 'bait ball'. There are thousands of them, all gradually in the process of being depleted as other creatures make big inroads while feeding on them.

Apart from sharks, among the most voracious guzzlers are thousands of common dolphins (*Delphinus delphis*) that move in from deeper waters for the duration. These graceful, streamlined creatures sometimes form 'super pods' that can be anything from a few dozen to thousands strong. Notably, the dolphins have developed a unique technique of sometimes isolating a shoal of sardines and actually 'herding' them in one direction or another, using streams of their bubbles, sonar and incredible teamwork to corral them into tight pockets. What has also been observed is that dolphins have the ability to give birth to their young shortly before the annual sardine run. This allows the adults to wean and teach their offspring hunting tactics during this incredibly food-rich season.

For the international community, it is the sharks that are the biggest attraction, in part because all dives on the 'bait balls' are without the protection of steel cages.

Divers who follow the sardine run encounter just about every species of shark in the Indo-Pacific Basin, as well as some from the Atlantic. These include the occasional great white shark, hammerheads, tiger sharks, zambezi sharks, spinners, grey reefs, ridgebacks, blacktips, and, now and again, a mako. The most common shark seen during this time are copper or bronze whaler sharks, as well as duskies.

There are times when dolphins and sharks can actually be seen to be working together in containing a specific 'bait ball'. They will take turns in plunging into the food source, mouths agape and gulping down what seems to be bucketsful of sardines. The most effective tactic here is to drive the 'bait ball' to the surface, which blocks off most avenues of escape.

In turn, this allows a variety of oceanic bird species to take advantage. Their numbers include the albatross, terns, Cape cormorants and skuas, together with scores of others. The most common winged variety is invariably the gannet, the ultimate scrounger of the ocean.

What quickly becomes apparent is that these birds have astonishing eyesight. They are able to plummet 30 or 40 metres into the frenzy of the 'bait ball' and, as we were able to observe as divers, sometimes reach depths of ten metres or more. Once underwater, they would swim around and animatedly snap and swallow just about everything edible within reach.

Then came the whales and there were a lot of them. Their numbers included Brydes whales which would feed by sounding and rocketing up from the depths with their great mouths wide open to engulf every sardine that came within range. We would film these monsters while diving and it was always an exciting experience. The migration also attracts some of the larger mammals like humpback and Southern Right whales, some that travel from the icy conditions in the Antarctic to the warmer waters of Mozambique and Tanzania for calving.

LESLEY ROCHAT
portfolio

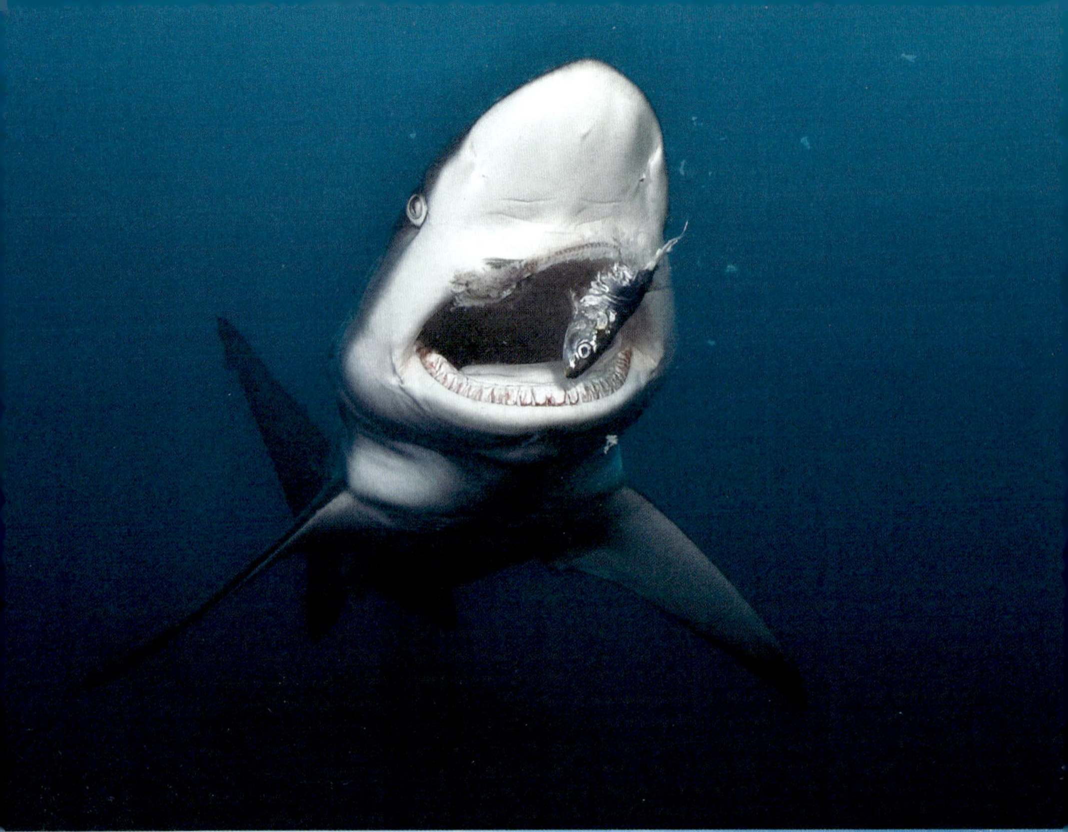

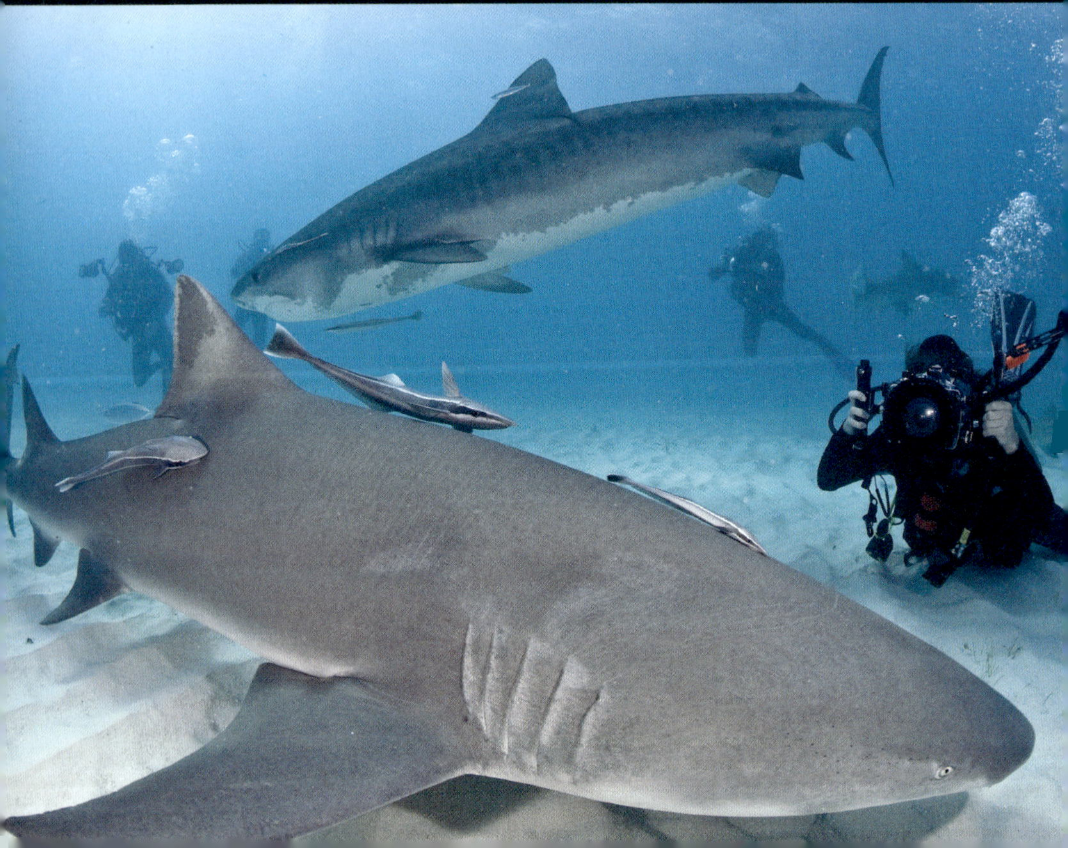

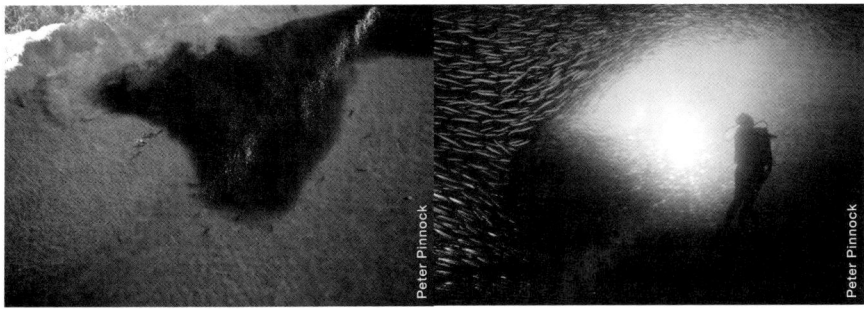

As the migration reaches the seas off KwaZulu-Natal, predatory tactics tend to change as more shallows are encountered. Around here, the continental shelf again heads out to deeper waters, which results in the sharks, porpoises and other creatures tending to herd the 'bait balls' into a variety of strategic bays and river inlets. These actions usually compact the fish and results in a blocking off of most avenues of escape.

Who said that some of the larger oceanic creatures cannot think?

Prior to his first 'bait ball' dive, Peter Pinnock — scuba tank strapped onto his back and ready to fall backwards into the water — thought: 'I've come this far, there is simply no turning back.'

'Just then, I knew, scattered about on the reefs along the rocky coast, lay the wrecks of hundreds of ships that had stranded along these shores, often violently and sometimes with a terrible loss of life. At one point we sped past the rusting hull of a ship that had been deposited by heavy seas high on the rocks *above* the waterline.

'The ocean around us, in contrast, bustled with marine activity. There were schools of dolphins everywhere, some in deeper waters, others surfing the waves and several drifting in and out of our bow wave. It was from above that we detected ornithological evidence of shoals of sardines nearby: a continuous stream of Cape gannets circled and headed north with us.'

A sense of expectancy among everybody on the boat became tangible as one of nature's great mysteries was about to unfold.

Descending into the sea, and sometimes directly onto a shoal of sardines for the first time, maintains Peter Pinnock, must be the touchstone of any underwater enthusiast's life. For him, he declared, it offered the best and worst moments of a diving career that spanned decades.

For a start, the uncertainty was sobering. So was its volatility, as this dense throng of marine life, often many metres wide, will — completely without warning — swing this way or that as predators

Paolo Fossati

moved in for the kill. 'Once you're up close enough to be embraced by that vibrant, pulsating shoal, there's no telling what's likely to happen next.'

'Absolutely nothing is constant when the sardines are on the move. One moment they're off to starboard, the next you're left blind and almost incapacitated as the fish swarm close around you, occasionally even leaving you in a string of grey shadows where it was difficult to tell up from down.

'Then the 'bait ball' moves on again and you emerge entranced and sometimes, a little fearful: there are predators everywhere. Meantime, the shoals head north at a steady pace to nowhere to keep their secret rendezvous.

'That cycle perpetually repeated itself once us divers were in there with them. A wall of fish would look impenetrable one moment and, in a flash, would split up ahead of us and reform again the next, often enough a few metres behind. Or the shoal might move left ten or 20 paces and then double back again towards the right. In-between, there were creatures that were feeding on them that cut swathes through this almost unbroken synchronisation of movement that we knew had been going on for weeks, months even.

'Having been totally enveloped by one shoal, I dropped down a few metres and held my breath. A myriad of sardines swarmed around and over me and suddenly day became night. Breathing out, our bubbles seemed to carve a path through this mass that was both enigmatic and utterly unpredictable. Then, in an instant, daylight would return.

'One of the most distinctive images with which I was left was that of almost total silence, which made the echo-locating clicks of the dolphins in the vicinity that more distinct. Quite suddenly, out of nowhere, a bronze whaler shark flashed over my head. Jaws open, it slammed hard into a solid wall of sardines. The shoal split up as if hit by a blast and the predator vanished into the morass. It took an instant for the sardines to close gaps again and continue on their journey.

'There was no question that there were a lot of sharks in the water. They were everywhere. At other times, there wasn't a single one to be seen. A shark might be feeding, jaws agape, a few metres away and we'd miss it. A second later a dozen blacktip reef sharks would whiz by and we weren't even spared a glance, even though one of the larger members of the pack swam straight through my legs, its dorsal fin bruising my thigh.

'Then, like phantoms out of the gloom, appeared another school of dolphins. One of these beautiful creatures that epitomised grace, glided straight toward me. It was clearly curious about these strange, black-clad fellow mammals that emitted bubbles. For a second or two it inspected me, gulped down half a dozen sardines and then moved away again as the rest of its pod indulged.'

Tell anyone that you've been diving in the sardines and the first reaction is inevitably: 'What about the sharks?'

A member of our team was French national Didier Noirot, who'd filled the slot as director of photography for Jacques-Yves Cousteau for a dozen years. Didier must be one of the most experienced divers on the planet since there are few open stretches of water in any ocean that, with a scuba tank on his back, he has not investigated. Asked if he was concerned about being attacked, he didn't hesitate: 'Sharks! I am not worried about the sharks. They're too interested in the sardines to eat me.'

It is also telling that during the sardine run, shark nets along the entire KwaZulu-Natal coast are lifted. This is a decision that was made immediately after the board was founded, and done specifically to spare large numbers of larger predators from being trapped in them. It's a truism, too, that shark nets do not discriminate between dolphins or sharks, or for that matter, turtles.

As a consequence, when Vic Peddemors dived with us, he was in the process of researching ways of preventing bottlenose and humpback dolphins — these species actively frequent the region — from becoming entangled. When that happens, he conceded, they drown.

Some very large packs of sharks follow the run of the sardines for its duration each year, tailing the shoals for what is thought to be a thousand kilometres or more. Most are bronze whalers, which, in adverse circumstances, are potentially dangerous to man.

Interestingly, the last shark attack that was recorded during the mid-year sardine run (before the German who was 'hit' while with Walter Bernardis) was in 1964 along this same stretch of coastline. It happened off Port Edward and, peculiarly, there was no evidence that sardines were in the area.

Peddemors: 'Historically, swimming was always banned whenever the board lifted its nets. Now the majority of local authorities use a system of what they like to call preferred discretionary bathing. This allows the nets to be kept out of the water for longer, while still allowing the occasional die-hard swimming enthusiast to use own discretion whether to enter the water or not.'

Information on Debbie Smith's company Diving With Sharks can be found on www.divingwithsharks.co.za.

Peter Pinnock's photos of South Africa's sardine run have been viewed around the globe. As with his other underwater photographs, many are world class and are included in the collections of numerous diving enthusiasts, clubs and collections, both in this country and abroad.

Examples of his craft — as well as his contact details — can be viewed at www.peterpinnock.com.

CHAPTER TWO

EXPERIENCED BLUE WATER DIVER

Most experienced divers can tell whether their buddies have had a shark encounter or two – and how things went – almost always by how they react in the presence of these predators.

That's the view of veteran salvage diver and former East London resident Peter Sachs. Navy-trained, he has dived on all sorts of wrecks under just about every condition imaginable, with and without sharks. He's been at it for years and is still on the job. These days, though, Peter is based in Vancouver, British Columbia.

Though Peter Sachs has evolved into an accomplished cold-water-fundi, his views are pertinent. His southern African experiences are of particular interest.

As he says, 'I suppose that sharks could be equated a bit like being stung by a bee. The more it happens, the less immune you become.

'I was far more relaxed in the company of sharks at the beginning of my diving career than I am now. These days, when I encounter a shark underwater, I watch it carefully. In the old days it would be there, but I'd only glance at it intermittently,' he told me. 'Obviously,' he added, 'things were a lot different whenever a shark indicated undue interest. It then becomes an all-systems-alert!'

Tommy Botha and Erik Lombard, both Springbok spearfishermen in their day and with their lifetimes spent in the water working as divers and as spearos – are the two individuals that Sachs regards as the most knowledgeable about these predators. He refers to them as 'sharkwise', and for very good reasons. Each has faced-off a handful of big whites while spearfishing in Cape waters and their approach – calculated, cautious and rarely bold or reckless — is instructive.

One encounter that Peter shared with Erik and Tommy was while they were searching for an ancient Portuguese shipwreck near the famed Chagos Archipelago deep in the Indian Ocean. It lies north of Mauritius and is said to have legendary links to pirates, privateers and other maritime adventurers, some of whom might have buried treasure there.

Certainly, there are a lot of shipwrecks in those waters. Also, the Port Louis Government is mindful about whom it allows to scratch around in those waters. Peter takes up the story:

Our yacht was anchored in a very remote area which has seen quite a few ships come to grief over the centuries: there were old cannon and other detritus strewn around on several reefs. The situation, remote as it was, was cause for cold comfort because we were operating far from any kind of assistance should we have needed it. Had one of us developed a bend, I suppose that would have been it. Fortunately, most of the diving was in shallow water.

There were other problems, like stone fish, sting rays on every single dive, not that they're a problem if you let them be and, of course, more sharks than anybody could have dreamt of. Some really big ones too ...

Then, after a series of storms, dawn brought with it one of those exquisite tropical mornings. A light breeze from the southeast was just enough to take the weight off the normally oppressive equatorial sun. From the surface we could clearly see the reef which ran in ragged lines ten or more metres below us.

We were just about to start our search and the guys were hyped, which was normal for the kind of enthusiasm that always presages the unexpected. Things were buzzing as we scrambled to don our gear and Erik was first into the water.

I was about to join him when I noticed that he made a rather spectacular movement that could best be described as a backward flip. The next moment he was back on the dive platform. He'd obviously seen something that spooked him and I could see from his demeanour that he stood back far enough so that not even the tips of his fins were touching the water.

'There's a ten footer right under the boat!' he spluttered, pointing downwards. Then he took a deep breath and said something about almost having bumped into a monstrosity of a tiger shark the moment he entered the sea.

Erik was of the opinion that this particular shark might have been attracted to the noise and bustle made by the crew on board, and also the clinking of the yacht's anchor chain when it was slipped. We'd just arrived and there was obviously quite a lot happening on board, including us firming up on the reef. For me, just then, the sensible thing might have been to find another spot to dive, or even better, go for a long walk on the beach.

Tommy and Erik had their own ideas. This was an opportunity, they declared, and with that they prepared to go in and hunt the damn thing. They were right, of course, especially if we were to do any constructive work that day.

Tommy picked up a small pouch from his dive bag and removed half a dozen elongated tubes, each about three inches long. In turn, they were opened and a high-calibre bullet inserted into the chamber. I watched closely as he went about it and noticed that the open-end formed a sleeve, which he slipped over the sharp end of his speargun.

He didn't need to explain that there would be an explosive impact the moment the spear struck the shark, usually on its head. In effect, the Powerhead – as it was called – created an improvised and reasonably effective firearm out of a speargun. With time, I was also to learn that Powerheads have been responsible for the deaths of an awful lot of sharks over the years, almost all of them aggressive, though others might simply have been curious.

Powerheads perform a useful role for divers within the shark's domain. Shooting a hostile predator with the kind of conventional steel spear normally used to bag fish is rarely enough to kill these beasts, especially if large. More than likely, such action could result in infuriating it, especially if the target is in the three- or four-metre range.

On the other hand, the effect of a shark being hit almost anywhere in the head by a high-calibre bullet usually results in death, if not immediate, then after a short while. The blast quite often forces the shark into a sharp spiral downwards, which divers refer to as the 'death dive'.

Apparently it is not so much the bullet that does the damage. It's the blast that does the trick, an explosion that releases a deadly and usually terminal volume of gases that neutralises the shark's central nervous system, and very effectively as well.

Tommy stuck two spare 'bullet tubes' under the cuff of his wetsuit and handed the rest to Erik, who'd armed himself the same way. They gave each other a quick nod and then slipped into the water. I stayed on board and offered moral support.

At first the pair moved towards the anchor chain, using the shadowed, starboard side of the yacht for cover. As soon as they'd progressed beyond the bow they went back-to-back. From where I observed on deck, it was clear they'd done this sort of thing often before because they moved about in the water in a carefully choreographed and synchronised pattern. A short while later they were back on board. The shark had disappeared, was their conclusion.

Whether by coincidence or design in the dives that followed later, I was eventually the one that ended up swimming along the outside track that day. For much of the time I worked the edge of the coral shelf where the reef dropped off hundreds of metres. In truth, I could easily have swum over Captain Kidd's entire treasure cache and noticed nothing, because throughout, I kept both eyes firmly fixed on the nearest deep, dark stretch of water.

Having spent most of my diving career in the East London area and Transkei, I've seen sharks aplenty underwater. It was early in my career that I had probably my most precarious experience, this one with a great white of impressive proportions.

Sean Mitchley, a commercial diver and in his day a spearfisherman of note, had recently joined me to form a salvage partnership that was working out quite well. Our main objective was to search for shipwrecks and we went on to make some important new discoveries. The most notable was the Portuguese carrack, *Nossa Senhora da Atalaia do Pinheiro*. Naturally it wasn't all work and no play. The sea was ours to enjoy and I used to do quite a bit of spearfishing for fun in those days. For Sean, that was his main sport: he was a regular member of

One of more than 20 bronze cannon recovered by Peter Sachs from the wreck of the Portuguese East Indiaman *Nossa Senhora da Atalaia do Pinheiro* which sank near East London in 1647.

the Border provincial team at that stage already and had competed at Springbok trials at various coastal venues.

There were many sites in the Eastern Cape that we dived, and for those who appreciate East London as a diving venue, you would almost certainly have heard of the infamous Three Sisters Reef. Quite imposing as diving sites go, it lies almost two kilometres offshore and hosts a variety of marine life. To dive there, however, one also has to cope with the elements because the weather, the sea and prevailing currents which can sometimes change by the hour, are often treacherous.

Spearos talk about the place with a certain reverence. They are all aware that the Three Sisters Reef has impressive drop-offs, caves, pinnacles that reach straight up from the ocean floor as well as a myriad of deep spots that attract game fish. There are also a few shallow areas. In fact, the 'Sisters', while not catering for all levels of diving proficiency, sometimes even allows novices to bag something interesting.

The reef has garnered a reputation over the years and most enthusiasts regard it as a diving venue that can be dodgy. As one Springbok spearo commented: 'That's certainly some reef – you never know exactly who might end up on the food chain ...'

Like most deepwater tracts along southern Africa's east coast, the 'Sisters' and its diverse sea life attracts a good selection of sharks of all the species, great whites included. Among the more prominent are packs of resident raggedtooth sharks which, in these colder waters are much more aggressive than those found in the almost tepid seas of Umkomaas' Aliwal Shoal.

Fairly large as sharks go, these critters tend to cruise many of the reef's gullies. Generally, they are pretty docile and some spearos like to shrug off the average raggie as being non-threatening. But I've seen both sides of a temperament that can sometimes be pretty volatile. I've always had the conviction that any shark three to five times your size needs to be taken seriously. Moreover, it is you, the diver, that is the interloper. The shark, in contrast, is in its own element. Even an accidental bite could cause serious problems because the average raggedtooth shark sports a brace of teeth that is about as intimidating and primeval as it gets.

More than once I've hauled myself out of the water at speed after being chased by a large raggie. Once I cleared the side of my boat in a single fluid motion that an Olympic pole-vaulter might have been proud of, my tanks, weight belt and the rest still strapped to my body.

My most memorable shark incident at the Three Sisters Reef didn't involve raggedtooth sharks. Rather, a great white was involved.

Sean and I had been driving back from Cintsa Bay on one of those lovely days that make living in this part of South Africa such a pleasure. We'd been working on the wreck site of the *Atalaia*, and since it was only a short detour from our route home, we decided to take a quick look at what was going on at Bonza Bay, from where the 'Sisters' are usually reached.

Water conditions were perfect. The land breeze had dropped and from the shore we could see the ocean wallowing lazily over the reef in the distance. The prospect of a fish *braai* that evening actually determined the move and we made the unusual decision to swim out to the reef from the shore. With us both being fit, navy-trained divers and accustomed to swimming long distances, it would be a cinch. Besides, there wasn't enough time left to still launch our ski-boat, which, we decided afterwards, was perhaps a bit of skewered reasoning on both our parts.

I drove my beach buggy to the area opposite the reef and we were into the water within minutes: it was a quick change and we were on the hop! Even better, once afloat, we discovered that the visibility was excellent. It would be a fine end to a productive day, we told each other when we stopped for a few moments and caught our breath beyond the breakers.

Little happened on the way out as we headed out towards deeper water. The reef en route, was unremarkable, though we did notice that there were a few gaps below that might have offered yellowbelly cod or bronze bream. Time was limited though and we sought a bigger prize. In any event, we were targeting the outer reef where the big fish lurked.

It took us about 40 minutes to reach the first of the three rocks for which we'd been headed. The 'Sisters' were pristine that afternoon, lazy in relatively calm conditions, but with quite a surge along the bottom as the swells moved over the reef. The rigours of the long haul out were soon forgotten as we drifted across the main fishing area, the sea alive with the 'clicking' sounds of the many organisms actively feeding on the bottom.

My first dive took me down and into a gully around 15 metres deep. While we spotted a number of 'keepers' within our range, I decided to wait for something bigger to come along.

That something 'big' did arrive a short while later: it was a huge raggie. By the time I surfaced, Sean was down a little further out and he called me over after he'd surfaced.

'Lots of raggies,' he said, which was when he suggested that we stick together and dive alternatively.

What this meant, basically, was that one of us would remain on the surface and observe the other dive while spearfishing below. It's an old spearo ploy, a comforting extra pair of eyes keeping watch for the unexpected. By now, I was a little uneasy about the way that the sharks were reacting towards us and very much aware that a 40-minute swim still lay between us and the safety of the shore.

Undeterred, Sean kept up with the hunt. There were a number of prospects around within the two- or three-kilogram range, but we'd formulated a policy long before of not shooting anything 'small' in the first half-hour of the dive. If there were big fish around, that might scare them off. We were also aware that with reef fish, the bigger they get, the more timid and illusive they were: perhaps that's the reason they grow into big fish.

Then it was my turn. I went down perhaps 15 metres and slid into a gully that backed up against the reef. It was a good feeling, always nice to cover your back. I'd been down about 90 seconds when a black steenbras weighing about ten kilograms swam into view, which I thought would do rather nicely. The presence of sharks at that moment completely eluded me.

The fish sensed my presence and moved in closer to investigate. Slowly, ever so slowly, I moved the tip of my spear towards him but it remained tantalisingly out of range. By now I'd been on the bottom for about two minutes and I needed to surface.

Somehow it was almost as if the steenbras was fooling with me and it kept its distance. Just before I headed for the surface, I took a shot anyway and predictably the spear fell short. By the time I'd taken my first glorious breath, the fish had vanished. Sean watched it all from above and I could tell that he was not amused.

Then it was Sean's turn again. I was still breathing heavily and my head hurt. Having hyperventilated on the surface, he was just about to drop his head to go down when I saw something large move from the bottom. This was no raggie!

The shark, which I could now see was a great white, was moving so fast that I didn't even have time to scream a warning. For a moment or two I just kept pace with its movement, amazed that something so large could move at such speed. At that moment, just as the predator moved in for the kill, Sean turned his head and saw it, jaws agape, coming straight at him. He braced up and instinctively thrust his speargun towards it.

A moment later a four-metre great white smashed into Sean and he was lifted clean out of the water. My immediate reaction was that my dive buddy had been taken, the two of them still attached and spinning around in a melee of limbs, fins and bubbles.

And then, in a blink of an eye, it was all over. The shark vanished into the dark blue murk beyond and I finned as fast as I could towards Sean.

Sean was in shock. So was I. We did some peremptory checks and found that the only damage was to Sean's spear. It was bent right over, the point facing backwards at him, the shooter. In a sense it was almost comical, but we knew we were in trouble because the brute could return.

'Let's go back-to-back,' I shouted at him. At that moment I thought that it would have been futile to try to race for shore. We took up our positions and braced ourselves for a second attack, both of us carefully scanning the direction in which I last saw the shark heading.

I spent a short while like this when I called to Sean: 'Looks like its gone.' There was no answer, which was when I turned round and saw my dive buddy swimming like an Olympic aspirant for the beach – alone!

Of course, I felt just great to be left hanging out on the reef all by myself. The fact that my partner had a hundred metres start on me didn't help either. Which was when I started with what I now regard as the longest single marathon swim of my life. It was also to be the loneliest.

It was a long haul back and I would fin five or six times and then turn and look back in the direction of the reef. Each time I did so I kind of half-expected to see something move in on me. Because of these antics, the swim took a good deal longer because I'd spent so much of the time checking behind me.

Eventually I made it. Sean was already on the beach, admiring his bent spear by the time I scrambled in.

We often talked about that dive in the days that followed. What we both regarded as remarkable was that Sean had managed to get his spear down the throat of the shark in that final split second before it actually struck him. That move, more by luck than design, almost certainly thwarted a fatal attack.

We were aware too, that even a relatively small injury inflicted out there might have resulted in serious consequences. Besides that, the predator would have had plenty of time to finish him (or even both of us) while we were trying to make it back to shore.

I don't have to add that that was the last time I ever swam out to

Three Sisters Reef, which is a pity, because it's one of the best dives in the entire region.

East London's Nahoon beachfront area has been the scene of some nasty shark attacks in recent years. Almost without exception, the attackers responsible were great whites that had targeted surfers or swimmers.

One theory is that these sharks migrate to the Natal coast with the winter sardine run and then return to southern waters when it's all over. This would explain why there seems to be a concentration of shark activity in the southern hemisphere winter months, around July and August.

For all that, I have dived or surfed Nahoon Reef alone many, many times. Some would regard that habit as a bad or dangerous choice, yet my preference has always been to dive alone, even if there is a huge advantage to having somebody in the water with you. As the saying goes – and I think it was originally coined by Jacques-Yves Cousteau: 'Dive Alone: Die Alone!'

The second issue is one of arithmetic. By having another person diving with you, the odds of being attacked are neatly halved. They improve considerably if your dive partner is not a great swimmer.

I make these observations because shark attacks involving more than one individual are rare. In fact one of the few known occurrences, worldwide, where this has happened was at Nahoon Reef. Two surfers were attacked by one shark. One victim was seriously mauled and needed about 400 stitches in his leg. The other died following the attack.

During the early 1980s there was a spate of shark encounters at Nahoon. There was nobody seriously injured, but there were a number of near misses. Chunks were bitten out of surfboards and paddle skis and, looking back, this could easily have happened to the surfers sitting on their boards at the time.

The shark scare went on for a while and we know now that the East London City Council was concerned. Shark attacks tend to affect tourism, which was when this august body decided to act. The prevailing wisdom was that there was one rogue shark out there and that it had to be caught and killed. Today it would be a crime to do this as great whites are protected.

The municipal beach manager was put in charge of the 'hunt' and I recall the disbelief when we heard that these bureaucrats – the new breed of shark hunters – actually planned to hang horse meat on hooks off Nahoon Reef in the hope of catching the 'lone' predator.

They set their baited hooks on buoys a couple of metres off Nahoon Point and all of us waited, though for some reason, nothing was communicated to a number of local sailboarders. When these fellows spotted a row of buoys out to sea, they used them as a gybe mark and since some enthusiasts were better than others, a number were unsuccessful at completing the gybe manoeuvre, which, for the novice, can be complicated.

Some sailboarders actually fell into the water and were almost impaled onto the baited hooks, not to mention any sharks that might have been attracted to the area. There would be a lot of kicking and cursing before the sailors would again make the safety of their sailboards.

Ultimately, East London's beach management was a lot more successful at catching sharks than some of us might have imagined. They actually ended up hooking an impressive great white after it had swallowed a section of equine hindquarter.

With the big shark snagged, it was the turn of the South African Navy to take a hand. They eventually got a line around the shark's tail and dragged it behind a patrol boat. By the time they got it back to harbour the shark appeared to be dead. It was certainly comatose.

The great white was loaded onto a bakkie near the ski-boat club slipway where a small crowd had gathered to see it, and because I was doing a surface supply diving course at the Port Rex navy base at the time, I was among those present. Since everybody was talking about the monster, I was keen to have a look at this predator for myself. So I ambled down to the ski-boat club.

By the time I got there the crowd had thinned and the shark was lying half on its back with its stark white belly exposed. Its tail had been shoved up hard against the cab of the vehicle while the head protruded almost a metre past the open tailgate. I was pretty impressed: this was the first time I'd been so close to a great white shark.

The fact that it was dead did not detract from its almost awesome presence. Looking at one of the most powerful and perfectly evolved animals on earth, the creature projected an atavistic brutality that both frightened and transfixed me. Somehow, I felt this great shark was the embodiment of all that is good and evil, of life and death. But I digress.

One of the first things I noticed when I got to the site was the shark's thickened, mackerel-like tail. If I were looking at a machine, this was its powerful engine. My gaze shifted slowly along the outer contour of its body: the perfect foil and symmetry of its almost too long, pectoral fins.

By this time I was alone with this dead shark. Finally I found myself staring at the mouth. It lay half-open and limp. There were flecks of blood coagulated to one side where the hooks had caused damage. I was transfixed by this dreadful maw and reached out slowly to touch its teeth, jagged rows of them, one layer over another. For a little while I ran my fingers over those sharp points, testing the prickly serrations. Each tooth, I found, was fractionally loose and seemed to be sprung in such a way as to saw through bone, which is one of the reasons why a great white can sometimes bite a large seal in half. I looked up and noticed a few stragglers talking about 20 metres away.

At that moment, it was perhaps a sixth sense that made me feel a little uncomfortable, which was when I withdrew my hand from the shark's jaws.

What immediately followed sounded like a rifle shot. The shark had just taken one final deadly snap and closed its jaws. I looked up at my hand and then back at the shark's head again, which was when I turned towards the stragglers and they had gone totally silent. Every one of them had seen what had happened and were looking at me, some of them aghast.

Nobody said a word.

I turned my back on the shark and walked away, well aware that the great white shark has 300 teeth in seven rows and that it can bite with a pressure of three tons ...

EPILOGUE

Sean Mitchley was killed in a motorcar accident near Grahamstown not long afterwards. His death highlights the fact that in South Africa you are far more likely to die on the country's roads than as a result of a shark attack.

The great white shark on the bakkie was taken to the East London Museum where the master taxidermist Greg Brett made a fine replica of it. It is still on display. This particular specimen was not responsible for the attacks at Nahoon Reef: its teeth imprints didn't match those found on the surfboards.

The East London beach authority hung more horse meat off Nahoon and caught another great white shark. Its teeth did not match up either. With the 'one rogue shark' theory now discredited, the East London City Council stopped its fishing antics.

I myself still prefer to dive alone. But I now do it in the Pacific Northwest near Vancouver, Canada. I sometimes even spearfish in Howe Sound or the Strait of Georgia. There are no sharks in this part of the world to speak of, that would attack humans. There are times, however, when I catch a glimpse of a seal or the shadow of something moving in mid-water. Then my heart jumps and my pulse rate will increase for a little while, at least until I have regained my composure.

The recovery of bronze cannon by Peter Sachs from the wreck of the Portuguese East Indiaman *Nossa Senhora da Atalaia do Pinheiro* can be found in Chapter Five of *Dive South Africa*, by Al J. Venter (LAPA Publishers, Pretoria). It makes for a fascinating real-life drama.

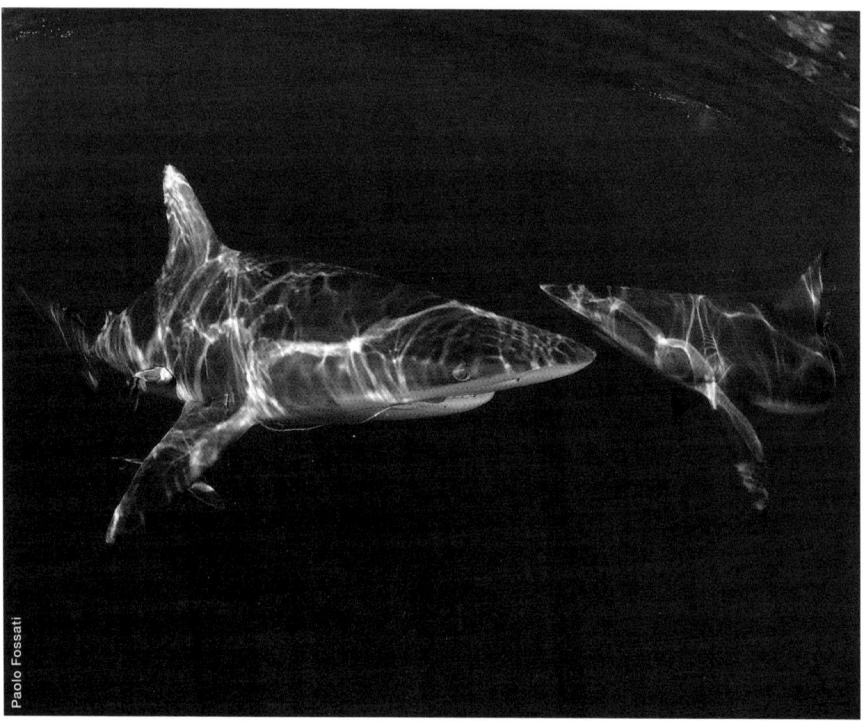

Paolo Fossati

CHAPTER THREE

THE BULL SHARK: MORE THREATENED THAN THREATENING

Geremy Cliff, one of the doyens at the Sharks Board, has a face that is arguably better known abroad than in his home country. He has featured as an authority on sharks in any number of films, including some notable documentary productions from the BBC, *Discovery Channel* in both the United States and Britain, together with several more under the auspices of Washington's *National Geographic*.

The bull, or zambezi shark — or more specifically, to give it the full scientific treatment *Carcharhinus leucas* — has not attracted anywhere near the public interest that is accorded either the great white or tiger shark. Yet all three species have the notoriety, as news reports indicate, for their predilection for attacking humans.

One possible explanation for this differentiation in status might be the bull shark's lack of size. Clocking in at between two and three metres on average, that makes it a lightweight compared to the largest recorded white shark which is in excess of seven metres and over three tons on the scales. The tiger shark attains a similar size, with one gigantic female caught off Indo-China in 1957 reported to be more than seven metres (almost 24 feet long and weighing in at roughly three tons – more than 6,000 pounds).

By comparison, the largest bull or zambezi shark from South African waters was, until fairly recently, a mere 2.9 metres and 350 kilograms, though early in 2009 a female bull shark of more than four metres was caught by an angler in the South Cape. Generally, the discrepancy in size is not matched by lack of aggression.

The bull shark does not have the formidably striking features of the great white and the tiger, but it can be recognised by a combination of stocky build, blunt snout and a terrier-like tenacity. Its features have given rise to its name: in vague terms, it might resemble a bull, with its comparatively small eyes probably contributing to the appellation. By contrast, the great white has large eyes.

At the outset, one may justifiably ask why one should compare these three species, if only because they are so very different physically and in temperament. The answer lies in the fact that this trio of ocean-going predators has been responsible for most of the serious attacks on humans in coastal waters throughout the world.

Several areas appear to have been subjected to more shark attacks than others, the majority either from bull sharks or great whites, though there have recently been increasing numbers of tiger shark attacks. While great whites are also common in warm temperate waters — there is even one recorded great white attack in the Antarctic — the other two species frequent warm semi-tropical coastal waters.

After three decades as a shark biologist, I have become accustomed to the unexpected, but one event that will always stick in my mind was a phone call from a Natal Parks Board official in September 1998 to inform me that a suspected bull shark had been found dead in the Inyamithi Pan of the Ndumo Game Reserve. The reserve is at least 70 kilometres from the sea on South Africa's northeastern border with Mozambique, although it is only something like 20 metres above sea level.

The pan itself, which is roughly eight kilometres long and a kilometre wide, has a maximum depth of between two and five metres. I have never actually visited the reserve, but on examining a detailed topographic map, I noticed that the pan is fed by the Pongola River, a large river by South African standards. Apparently water levels in the Pongola River have fallen in recent years, which resulted in some bull sharks being trapped in the pan.

This specimen was already dead when it was spotted being pushed around by an inquisitive hippo. The shark was hastily dispatched to our laboratory in Umhlanga near Durban, where its identity was confirmed: a juvenile male bull shark of roughly two metres. There was no indication of any external injury, which might have caused the animal's demise.

Despite an abundance of fish in the pan, including the battling tiger fish, barbel (catfish) and tilapia – which would have provided this predator with plenty to eat – a dissection revealed that the stomach was empty. Liver weight is a good indicator of the condition of a shark, because this large organ is the primary storage site for the body's reserves of oils and fats.

In some sharks the liver can constitute as much as a third of the body weight. In a bull shark the expected liver weight of a 90 kilogram specimen would be something like eight kilograms. This particular individual had a liver that was almost 25 per cent heavier, much more than we expected.

This discovery should not have come as a total surprise, after all the bull shark is known for its ability to penetrate freshwater. It has been recorded in Lake Nicaragua and in numerous large river systems, including the Mississippi, Amazon, Tigris (Iraq), the Gambia River, the Ogoue river system in Gabon as well as the Swan and Brisbane Rivers in Australia. Such records not only confirm its liking for freshwater, but also reflect its widespread occurrence in coastal waters of all tropical and subtropical seas.

At one stage it was thought that the population in Central America's Lake Nicaragua had become trapped in the lake system and had therefore been forced to become totally independent of the sea. This is not the case and although some individuals may breed in the lake, most individuals leave the lake and utilise the brackish coastal waters. As an indication of their inclination to venture far inland away from the sea, bull sharks have been found 3,700 km from the sea in the Peruvian Amazon River. In the Ganges River they are probably responsible for the numerous attacks on Indian pilgrims who were formerly attributed to the little known and poorly studied Ganges shark *Glyphis gangeticus*.

Focusing on the southeast coast of Africa, these sharks have been found in the Zambezi River system up to 1,120 kilometres from the sea. Their presence has also been recorded in the Sabie and Limpopo rivers and many of the river and lake systems on the east coast of southern Africa.

Water temperature limits their distribution, as the species is rarely found in water much colder than 20 °C. These sharks are most common in KwaZulu-Natal coastal waters during the summer months. With the advent of winter and declining water temperatures, they generally move northwards towards the equator. The southern limit of their southern African distribution has traditionally been regarded as Cape St Francis in the Eastern Cape. In January 2009, shark researcher Meag McCord from the South African Shark Conservancy and her colleagues caught the largest bull shark on record, the four-metre female mentioned earlier in the sprawling, unspoilt Breede River estuary in the southern Cape.

This confirmed anecdotal reports of the presence of these sharks, but way back in the 1960s and 1970s. Subsequent research has indicated that this was not a one-off occurrence. This record has extended the southern range of this species by over 350 kilometres and is yet another example of how little we know of these magnificent animals.

One of the most important localities on the KwaZulu-Natal coast for this species has undoubtedly been Lake St Lucia, part of a World Heritage Site, aptly known as the Greater St Lucia Wetlands Park and now the iSimangaliso Wetlands Park. The lake system balances on something of a knife-edge in that evaporation of about 1.300 millimetres per annum is on par with the average annual rainfall. Hypersaline conditions occur during periods of drought when evaporation from the approximately 40,000 ha of water, with an average depth of only a metre, albeit below sea level, exceeds freshwater inflow, mainly from the Hluhluwe and Mkuzi rivers. As a result salinities as high as 110 parts per thousand (ppt) — (seawater has a salinity of 35 ppt — have been recorded.

The research of Dr John Bass and his colleagues from Durban's Oceanographic Research Institute in the 1960s and 1970s revealed that bull sharks not only tolerate freshwater, but are found in parts of the lake where the salinity is higher than 35 ppt. It would appear that once the salt content exceeds 50 ppt, the sharks leave the system and return to the sea.

Most of the bull sharks found in the lake are newborns and slightly larger juveniles, together with a few adult females. It would appear that the females enter the system to give birth in early summer. The newborns remain there during the early part of their lives. Larger sharks may move into the lake and then out to sea as they mature. Bass concluded that St Lucia is probably the most important bull shark nursery ground in South Africa.

At the time of writing, the mouth of St Lucia had been closed for almost a decade. In the past, the mouth was kept open by almost continuous dredging operations. A change in the management strategy has resulted in a situation where nature has been allowed to take its own course. This has unfortunately denied the bull sharks on the South African coast access to their most important pupping ground, presumably forcing them to use other major estuarine systems, not only in Mozambique, but also further south along the east coast of South Africa.

The average litter size for the bull shark is about nine (with a range of six to twelve); size at birth is just over half a metre and about 2.8 kilograms. Gestation is believed to be just less than a year and it is highly likely that the females have a two-year reproductive cycle. This would mean that a female would only produce nine young ones every two years. The year-long interval between pregnancies is common among sharks, attributed to the need for females to replace the considerable energy stores lost in the nourishment of their developing embryos.

Where is the best place for divers to view these magnificent beasts? Spearfishermen will report that it is not unusual to see solitary individuals when hunting the numerous near-shore locations.

My one unnerving encounter with this species was while testing an electrical shark repellent, the Shark POD, on the deep pinnacles at Ponta d'Ouro in southern Mozambique. On that memorable day, conditions were as close to diving in a natural aquarium as any diver could hope for.

The water was clear and warm and there was very little current. Schools of giant kingfish, big-eye kingfish and barracuda were milling about around us. All that was lacking were the sharks. Sod's law!

When one wants to spear fish without being harassed, sharks invariably appear, responding to the blood and thrashing of a speared fish.

Geremy Cliff at work and at play – 'breathing life' into a distressed shark before he releases it into open water. (Photo: Geremy Cliff)

On this particular day we wanted to attract the sharks to see if the electrical field of the Shark POD would repel them. My buddy, Rod Haestier, took a last deep breath and left the surface. He finned slowly and effortlessly down to about 15-metre depth where he stopped, picking out one of the tropical yellowtail slightly below him. I then saw him move forward, this time with a lot more purpose, and fire.

No sooner has his spear left the gun, when a bull shark shot out of the purple gloom directly towards his fins. I watched the scene unfolding below from the comfort of the surface. Rod had no way of knowing that there was a shark right behind him. Even if he had seen it, it was moving so fast that he would probably have been unable to take evasive action. Then, quite suddenly, the shark stopped cold, its snout almost touching Rod's fins. Fortunately, Rod remained stationary and, as he had missed his target, the shark banked and slowly swam off as though nothing had happened, with Rod blissfully unaware of his close encounter.

The firing of the speargun must have been a signal that the shark could not resist. One can only speculate what might have happened had Rod's spear hit home.

One less fortunate individual was a young spearfisherman who was diving on Aliwal Shoal on the KwaZulu-Natal South Coast. The water was extremely clear and about 22 metres deep. He shot a fish of about two kilograms on the reef, abandoned his gun and surfaced. About eight metres from the surface, a large bull shark seized his right calf from behind, shook his leg and then released it.

Clearly, the shark had been both attracted and aroused by the spearing of the fish and its resulting struggles, however, the diver's rapid departure for the surface proved a far more attractive signal to the shark than the wounded fish.

In my ongoing investigations of shark attack on the southern African coast I have come across more than one such incident where a spearfisherman has shot a fish and a zambezi shark has responded almost immediately by attacking not the struggling fish, but the diver himself. It appears as if the shark has responded in this manner because it resents the presence of another predator in its 'territorial domain' and chases it away.

These two incidents indicate that of all the recreational users of the sea, spearfishermen are at the greatest risk of shark attack, followed closely by surfers.

For scuba divers, apart from this pinnacle at Ponta d'Ouro, the locality that is best known for viewing bull sharks is on Protea Banks, a short boat ride off the popular KZN South Coast launch site of Shelly Beach. The reef is generally about 30 metres (up to 40 metres and more in places) deep and about eight kilometres long.

My first scuba dive there was in 1996. I had heard so much about the large "zambies" or "zams" (as divers and anglers like to refer to them), that I was both excited and apprehensive. My feelings were reinforced by our divemaster's descriptive and somewhat overdramatised pre-dive briefing. As we fell off the boat and the bubbles cleared around us, there they were, two rather large bull sharks swimming only about 10 metres below the surface. Unfortunately that was the end of our interaction with these magnificent animals as they soon disappeared.

On future dives I found that the sharks were generally rather timid and the tendency was to find them in much deeper water. Usually it was during our leisurely ascent that we saw the sharks rising slowly, albeit purposefully, off the bottom and heading towards us. Once they had satisfied their curiosity, which was usually done at a distance of at least ten metres, the sharks levelled off and slowly descended back to the depths.

Invariably, these specimens of about two to two and a half metres long, were accompanied by one or two remoras or shark suckers. These fish have a dorsal fin, which is modified to form a large sucking disc. The remora has evolved a lifestyle whereby it clings firmly to a large shark and, in so doing, hitches a free ride. This unique behaviour ensures that it is always present when food is available, as it feeds mainly on small pieces of fish that break away when the shark rips into its prey. Remoras are also an immediate indication that there are sharks in the water with you, the bigger the remoras, the bigger the sharks ...

Although these fish are quite capable of swimming on their own, they favour either sucking onto their host or gliding effortlessly alongside or below it, often relying on the slipstreaming effect to pull them through the water.

Sadly, something of an acrimonious relationship has developed over the years between scuba divers and some of the charter ski-boat fishing operators working Protea Banks. The former have accused the latter of unnecessary killing the very bull sharks that its scuba diving clients come to view. The latter customarily respond by accusing divers of deliberately trying to discredit their charter fishing operation. The dispute has had some serious consequences, with the fisher folk having taken the matter to the High Court in Pietermaritzburg, with consequent adverse coverage in the media.

The Judge ruled in favour of the divers, stating that although the pamphlets distributed by the divers referring to 'senselessly slaughtering' sharks were defamatory, they were nevertheless in the public interest and therefore not actionable. The conflict of interests did not stop there. Indeed, divers and other environmentally minded groups consistently argue that the bull shark needs more protection. Fishermen contend that what they are doing is not illegal and that there is no law to stop them targeting sharks or selling their shark catch as commercial anglers.

Some of the justification for killing sharks, the fishing community reckons, is that they are seen as the 'taxman' of the sea, taking a significant percentage of the fish hooked by the boat anglers.

Those who view the controversy from an economic perspective will contend that a shark is worth more alive on Protea Banks, where it can be viewed repeatedly by groups of scuba divers, than dead, simply for the price paid by a bounty hunter.

Internationally, one of the best-known venues for both bull sharks and shark viewing in general, is at Walker Cay on Abaco Island in the Bahamas. On one side of the island, huge numbers of Caribbean reef sharks and spinner sharks are attracted to feed on a 'chumsicle' (as in popsicle), a frozen mass of accumulated fish entrails frozen with a wire hawser.

The 'chumsicle' is deployed over a sandy patch of seabed, in water eight to ten metres deep and surrounded by reef and close to the open ocean. On arrival at the selected site, the boat engines are driven at high revolutions, which to the sharks in the area represent a dinner gong, something to which the sharks have been conditioned. On the other side of the island some of the fish waste is also discharged in water, where groups of large bull sharks jostle for the scraps. Snorkel divers watch in amazement as the sharks approach almost to within touching distance.

It was here that renowned shark researcher, Dr Erich Ritter, was demonstrating his hypothesis that sharks pay no attention to humans in clear water, even with an abundance of bait in the water to arouse the shark's appetite. Ritter claimed that he could protect himself by relaxing, thereby slowing down his heart rate.

In an article that featured this activity, the manager of Walker Cay Marina, Gary Adkinson stated that Erich was standing in chest-deep water with a television documentary presenter when his theory backfired and the researcher was attacked. Gary threw a chunk of fish into the water about five metres away from where they were positioned. A remora or shark sucker seized the bait and fled straight towards Erich. A bull shark also swimming near Erich then snatched at the bait and in so doing, bit directly into Erich's calf.

The resulting injuries required immediate evacuation to Palm Beach, Florida, where a team of specialists managed to save Erich's leg after an operation that lasted several hours.

These and other incidents have reinforced the bull shark's reputation as a highly aggressive species. Having said this, humans are not part of any shark's natural diet.

Perhaps one of the most fascinating aspects of shark biology is investigating the diet of these creatures. Despite their position at the top of the food chain, sharks do occasionally ingest some really unusual items. The tiger shark is exceptional in that it displays a complete lack of discrimination in its diet, ingesting a wide variety of both inedible and edible items. The bull shark is not as noteworthy in this regard.

In over two decades of investigating the diet of bull sharks caught in the shark nets deployed along the popular southern African recreational beaches, the only really unusual items that I have found while doing this work at the Sharks Board have been a single roast potato and on two occasions, plastic bags.

This database of 510 stomachs from sharks captured in the shark nets revealed that 201 were empty. There was an almost equal split between bony fish which were found in 55 per cent of stomachs with food and elasmobranchs (cartilaginous fish: sharks and rays), also with 50 per cent. This was followed by mammals at nine per

Photo: Gérard Soury, courtesy of Walter Bernardis

cent, comprising mainly dolphin remains, but including terrestrial remains from a rabbit, mole, dog and a domestic cat.

One of the more bizarre findings was that of two human feet, freshly ingested and avulsed at the ankles together with several chunks of human skin and associated muscle from the thigh region, all presumed to have been scavenged from a drowning victim.

Historically the bull shark was known in African waters as the slipway grey because of their habit of congregating around the whale-processing slipway in Durban harbour in the old days before whaling was banned.

The industry began in Durban in 1908 with the arrival of two Norwegian whalers and grew. At one stage there were six whaling companies operating out of Durban harbour. A factory was then built on the south side of the port and Durban became the world's largest shore-based whaling station.

In 1960 a total of 830 people were employed in the industry and catches peaked at 3,860 whales in 1965, until the industry closed in 1975. In latter years the gunner – usually the vessel's captain – fired a 50 kilogram explosive harpoon from the bows. This missile was attached to a wire cable, and if the whale was not killed instantly, the cable was hauled in and a second harpoon without a line would be fired into a vital spot. The dead whale was pumped with compressed air and marked with a buoyed flag and a radio beacon for later collection.

After the day's hunt, all the whales were collected and towed to port, where, tail-first, they were hauled out of the water onto special railway wagons on the slipway. This process attracted great white and bull sharks, and no doubt gave rise to the name by which the latter became known – slipway greys.

Bull sharks have a habit of making their presence felt all over the world, invariably for the wrong reasons.

Recife, a large city on the northeast coast of Brazil and capital of the state of Pernambuco, is one such locality. In 1992 the city was rocked by a succession of six shark attacks. Although none were fatal, the number was considered excessive, given that Pernambuco had only experienced three attacks in the previous seven decades.

Worse was to follow with a total of 20 attacks over the next two years, of which four were fatal. At a dedicated workshop to which I was invited at the end of 1995, we received reports that there had been a total of 34 attacks in six years, of which seven had been fatal.

Reasons for these alarmingly high statistics included a marked increase in the popularity of surfing, a general degradation in the near-shore marine environment through a variety of man-induced disruptions and the fact that potentially dangerous sharks such as the bull and the tiger occur in the region.

Perhaps the most important contributory factor was the establishment of the nearby port of Suape in 1992, which saw a major disruption to the highly productive mangrove-dominated estuarine environment. It was thought that this development had forced the sharks out of Suape and into either the coastal waters or the adjacent Jaboatão River, which opens into the sea at the beaches where the attacks have taken place.

The presence of a relatively deep channel running parallel to the coast and a relatively short distance offshore would have enabled the sharks to travel along the coast and then venture inshore where they would have come into contact with the large numbers of swimmers and surfers.

The fact that this water is warm throughout the year (as high as 26°C in winter) and also murky, meant that it was ideally suited to bull sharks. In an attempt to solve the problem, the authorities banned surfing and swimming at those beaches with the highest incidence of attacks. I was stunned to learn that an enormous number of surfboards had been confiscated. The owners were then given the option of paying a small monetary fine or having the board destroyed. In that region alone, there were something like 500 boards burned.

In some places at low tide the fringing reef does provide a natural barrier, but these areas are not favoured by the surfers because the reef not only acts as a barrier to the passage of sharks but to the swells as well. Consequently there are no waves within these areas.

Another recommendation of the workshop was to conduct research to confirm the relative abundance and any seasonality in the occurrence of sharks within the long-shore channel, with particular emphasis on bull and tiger sharks, which, given the severity of the incidents, were the most likely culprits.

Although some research has been completed, the shark attack problem persisted. At a follow-up workshop held in July 2004, I learned that 44 attacks had taken place in Pernambuco since 1992, of which 14 had been fatal. There were 25 attacks on surfers, of which only three were fatal. However, of the dozen or so swimmers attacked, there has been a staggering 75 per cent fatality rate. The statistics may be even higher in that there were a possible 20 additional cases in which scavenged corpses were discovered and there was no way of confirming that shark attack had been the cause of death, rather than drowning.

The truth is that Brazilians often don't bother to keep statistics in areas that are not regarded as tourist 'drawcards': the actual figures could actually be much higher.

In trying to resolve issues that surround the unpleasant spectre of shark attack, there are generally two schools of thought.

One group regards the sea as a marine wilderness and believes that by entering the sea, humans are invading the domain of the shark and should do so at their own risk. They would therefore recommend that the sharks be left alone and, if anything should be done, it must be to continue the ban on surfing and swimming.

Others, including most surfers, are frustrated at not being able to pursue their favourite pastime and believe that some action should be taken. In Recife I was closely questioned as to what I would recommend, given that the beaches of KwaZulu-Natal in South Africa and those in Queensland and New South Wales on the east coast of Australia are the only ones in the world protected by large-meshed nets that are designed to catch and kill dangerous sharks, thereby reducing the risk of attack.

While these shark nets have been spectacularly successful at reducing shark attack (in the case of Durban, by far the largest coastal city with the highest number of beach users, there has not been a serious shark attack since nets were introduced in 1952), I had to concede that these nets are not selective in that they not only catch the potentially dangerous sharks but also relatively harmless sharks such as small hammerheads and grey nurse (raggedtooth) sharks, as well as rays and to a lesser extent turtles and dolphins.

Baited hooks attached to large floats (termed drumlines in Queensland, Australia) where they are widely used, at least offer the prospect of reducing the number of bull sharks in the area, while posing little threat to other sharks and marine life in the region because of their large hooks. Based on their successful deployment in Queensland, these shark fishing devices were introduced at the popular tourist beaches in southern KwaZulu-Natal in 2007, where they fish alongside the shark nets, most of which have been deployed for nearly five decades.

In Recife meantime, the authorities have introduced a combination of drumlines and longlines, but no nets. These have virtually eliminated shark attack. About 80 per cent of the sharks caught are found alive and so they are transported out to deep water and tagged and released.

My own investigations into shark attack have taken me to Port St Johns quite often. Second Beach on this quaint Wild Coast holiday resort saw six fatal attacks between 2008 and 2012. We suspect that bull and possibly tiger sharks – rather than great whites – to have been the culprits.

Understandably, the local community is highly agitated and wants an end to this unpleasant trend. However, this is something that is not as simple as hiring a professional hunter to kill a vagrant lion which keeps straying out of the bush and coming into contact with humans. To date our research has confirmed what local, veteran rock and surf angler, Tony Oates, has found over decades of fishing – that the sea around Port St Johns is full of sharks. In the summer months he has often seen large bull sharks wallowing on the surface in the river mouth.

Moreover, the area not only attracts bull sharks, but is an important nursery ground for this species. At this stage it is mere speculation, but with the decade-long closure of St Lucia, the Umzimvubu River in the south may have attained even greater significance as a nursery ground than at any time in its history.

As part of the KZN Sharks Board's current research into the Port St Johns shark attack problem, we have caught six newborn and two one-year-old bull sharks in the river since March 2012. In all but two of the sharks, an acoustic tag was surgically inserted into the body cavity, which will enable us to monitor their movements by means of a series of listening stations deployed at regular intervals up the river.

Two large bull sharks (and hopefully more in the future) have been tagged in a similar manner and listening stations have also been deployed off local beaches. This is not going to eliminate shark attack, but it will give us a better understanding of bull shark movements in the area. With an enhanced understanding, we are consequently better able to mitigate the risk of shark attack.

As a group, sharks are under increasing threat and there is growing international concern over the impact of human activities, primarily fishing – both angling and netting – on their stocks.

Sharks and their cartilaginous relatives, the rays and skates, are characterised by slow growth rates, accompanied by late maturity and a low fecundity, making them far more vulnerable to overfishing than bony fish. Sharks Board researcher, Sabine Wintner has shown that bull sharks are particularly slow-growing, taking in the region of 20 years to reach maturity.

Perhaps a greater threat to the bull shark is the degradation of their estuarine habitat, which is essential for the survival of the newborns, as a result of development and poor land use practices. Water pollution, water extraction and siltation of our rivers through soil erosion and destruction of the flood plains are all too frequent.

In conclusion, one can but wonder what the future holds for this species in southern Africa waters. Bull sharks will continue to attack the occasional human but there is no doubt that man poses a greater threat to bull sharks than the other way round.

BULL SHARKS

In a more recent development involving bull sharks in August 2011, the Seychelles Archipelago, a favourite resort and diving destination in the Indian Ocean, was stunned by two fatal shark attacks. These took place 16 days apart and at the same location – Anse Lazio beach on the island of Praslin.

It was an enormous blow, especially since that beach and surrounding areas had no history of shark attack, never mind two in a row. Sadly, British newspapers went overboard because both victims were British. One of them on his honeymoon.

Within days, the Sharks Board had an urgent request from the Seychelles government to send somebody to look into the matter and possibly offer a solution. Which was why, only four days after the second attack, Mike Anderson-Reade and I arrived in Praslin.

What was tragic about the event, was that prior to our departure, I had been inundated by requests for comment from the media, mainly British. As a consequence, headlines in the English tabloids portrayed us as 'Shark Hunters from South Africa', a myth I tried my best to dispel. There was a reason, of course: immediately after the second attack, the government, desperate to prevent a third attack, sponsored a bunch of local fishermen to catch sharks in the immediate vicinity of the beach.

Seychelles officials in charge didn't have to be told that any chance of success was remote, and for several reasons. First, there was simply no proof that it was the same shark and, second, had the same predator been responsible for other attacks, it may already have left the area.

On arrival on the islands, we were shown several sharks that fishermen had caught, but these were mainly lemon, grey reef and blacktip reef sharks, none of which were capable of the two fatal attacks. Meantime, all water activities had been banned at Anse Lazio and adjacent beaches and I couldn't help feeling for the many tourists who had been deprived of something they had probably set their hearts on for some time. Those who didn't go after other pursuits were now forced to do little more than sit on the beach and watch the waves roll in.

We carefully examined the photographs of the injuries suffered by the two victims and both Mike and I were pretty certain that the culprit was a fairly large great white shark. But since we had nothing tangible with which to back our hypothesis, there was really no way of knowing whether the same creature might have been responsible.

A minute tooth fragment had been recovered from the second victim (who had lost his left hand with his wedding ring, presumably in trying to fend off the shark), but that needed something more intrusive than we had available on the island. Having examined the fragment, which had serrations, it was too small for us to identify the species.

As none of the beaches around Anse Lazio had any significant wave action – though seasonal change in monsoon winds would ultimately change this – Mike's recommendation was that the authorities insert exclusion nets at selected beaches. He felt that as an interim measure, the step would act as a physical barrier in keeping sharks out of a small swimming area and this would be achieved by floating a small mesh net out from the shore and back, ensuring that the net reached to the seabed. Using small mesh – as opposed to medium or large – would eliminate the possibility of sharks, turtles and large fish becoming entangled.

Mike had had considerable experience in this after heading the Sharks Board's contract with the Hong Kong government in the mid-1990s. As a consequence, 19 such nets were provided for these beaches following three fatal attacks.

On our final day on the island, some fishermen landed a 3.5-metre tiger shark. Given our belief that a tiger shark was responsible, we decided to check the contents of this creature's stomach. By now, a large contingent of onlookers had gathered around the fishing boat to catch a glimpse of 'the dreaded shark', so, in a bid to achieve privacy, we boarded the boat and headed offshore a short distance.

There we removed the entire stomach and placed it in a plastic bag before returning to the beach. Ashore again, we were escorted by the island's police chief to the local mortuary where we examined the stomach. We found the semi-digested remains of an entire hawksbill turtle and a single flipper from another hawksbill, as well as a noddy tern. There was no sign of any human remains, which was not unexpected, given that the first attack had taken place more than a fortnight earlier.

What we also searched for was the second victim's wedding ring, but in the end we found nothing. Had we done so, it might almost have been a fairy-tale ending, bringing closure for the honeymoon widow and providing reassurance for the tourism industry that the threat of another shark attack had disappeared. But it was not to be.

One of my recommendations to the Seychelles government was that they send the tiny tooth fragment recovered from the second attack victim to a marine laboratory in the United States for genetic analysis. This they did and the report that came back to me several weeks later was that the fragment was from a bull shark and not a tiger shark ...

Having launched their boat, a group of divers set out from the beach at Sodwana Bay, a favourite South African dive destination. Sharks are often spotted along some of the outlying reefs, but like other Indian Ocean destinations, their numbers have been decreased by Chinese longline fishing boats that plunder our in-shore waters. (Photo: Author)

CHAPTER FOUR

GREAT WHITE CAGE DIVING WITH MIKE RUTZEN

Mike Rutzen declared in *Sharkman*[1], a television documentary he made for *Discovery Channel*, 'that many people believe that the great white shark is a mindless killer, with only one desire – to hunt and feed on whatever it wants.' That is simply not true, Mike declared when we recently spent time together at Gansbaai, a small fishing village a couple of hours' drive east of Cape Town.

He said that in his pursuit of trying to understand these magnificent creatures, he'd discovered something totally unexpected while in the water with them. Some of the great whites could be astonishingly aggressive, he told me, but at the same time, there were others that were quite passive. Occasionally he'd encounter what he liked to term a 'player', which was when he suggested that there were specific sharks that didn't seem to mind a bit of interaction with people.

'You obviously have to look for them – you kind of test the waters, as it were,' he stated, adding that it also needed the kind of aggressive experiment that put himself and his team in danger. But it was the only way, he reckoned.

'At the end of it, I encountered a kind of secret non-aggressive side to them, which provided us with a totally different perspective to these so-called "killers ... " ' he declared with characteristic non-chalance.

It's symptomatic of the man – and what he does in waters when there are sharks present (and something that has been commented on by others who have worked with him, both above and below water) that he tends to give the impression of being totally detached from any kind of potential threat.

One American film producer declared that his mién fringed on mindlessness, and that Mike Rutzen couldn't care less about the potential dangers of diving with so-called man-eating sharks. It also says much that almost all the film crews who work with him while he does his thing with great whites, were put through their paces by him personally.

The truth is that Mike Rutzen cares very much what happens around him. Were that not the case, he argues, he would probably not be alive today. As he admitted after a few beers at Gansbaai's Birkenhead Restaurant, 'I just don't like to show emotion ... also, the nature of the work requires that I keep calm. If I indicate fear, the shark is aware of it, so I put such things out of my mind when I'm in the water with them ...'

In *Sharkman* and other films made for *Discovery Channel* and other international networks — all of which took many hours in the water while filming, sometimes in an extremely hostile environment — he told me that he was able to demonstrate this attribute once or twice by actually bringing one of these great creatures 'kind of' under his control. Mike referred to it as a form of hypnotism, and went on to explain that while there was an actual measure of acceptance of the diver by the shark, it was the predator that retained control, even if it allowed him to 'hitch a lift' on its pectoral fin and drag him along underwater for almost the length of a football field.

Tourists boarding a shark viewing boat moored alongside the quay at Kleinbaai; at any one time during the season there can be eight or nine boats off Dyer Island viewing great whites. (Photo: Author)

Mike Rutzen, dressed for work at his 'office', which is invariably in the sea with his beloved predators. (Photo: Author)

'It was a really strange feeling the first time I did it ... almost surreal ... the shark, a big one of between three and four metres, was aware that I was hanging on and that he was propelling both of us along ... and now and then I'd look down and I'd see this great big black eye looking at me.'

That experience, Mike concedes, was one of the touchstones of his life. 'It was both momentous and unforgettable,' was how he later described it.

'It was also pretty unique, because until then, there had been numerous experiences with great white sharks while I was free-diving with them in our South Cape waters. Throughout, we had to accept that the experiences were both risky and not always pleasant because there were moments when specific great whites – usually quite big ones – would become altogether too interested in the divers in the water. Then I would signal to the crew to head back to the boat and we wouldn't hang about ...'

But, he maintains, if you count the number of hours that he has spent in the water free-diving with great white sharks, then such occurrences are comparatively rare.

'Being among the largest creatures of the deep and at the top of their own food chain, great white sharks are invariably curious. At the same time, they are not naturally aggressive, far from it,' he maintains.

He points out a freak accident that occurred in April 2008, when one of the dive boats with almost two dozen passengers on board overturned and dumped everybody into the ocean. There were half a dozen other boats out there, all carrying passengers and each of them chumming the water with fish blood and offal in attempts to draw the sharks closer. In fact, there were already a number of large whites in the immediate vicinity.

When the boat was flipped over by an inordinately large swell – something that had never happened before in that maritime tourist environment – the sea was suddenly full of floundering, screaming and mostly helpless people. Of course, all the other boats came rushing over and almost everybody was rescued. Two Americans and a Norwegian died as a result of the accident: they were all trapped below deck and drowned.

'But it is significant that there was not a single attack on any of the humans involved ... not one, and there were 16 survivors hauled out of the sea. Shark specialist André Hartman commented afterwards that there weren't even any *attempted* attacks by sharks, even though it took the boats several minutes to pull everybody out of the water. 'And yet they were right there in the water with all these panicky survivors, no question about it ... they stayed away from what should have been pretty obvious targets,' he added.

'Though some of the boat crews thought at the time that there might be an attack, it never happened ... and that tells you a lot about these so-called killer creatures ...' were Mike's closing thoughts.

Caroline Castell-Elster, a diving enthusiast from Britain who has free-dived with packs of blacktip reef sharks as well as zambezi (bull) sharks south of Durban and several times among raggedtooth (grey nurse) sharks on Aliwal Shoal and off Sodwana Bay, reckons that her most memorable shark diving experience took place while she viewed great white sharks on board one of Mike Rutzen's boats.

'There was nothing "open water" about the experience,' she says. 'We were all mustered on deck and took turns to go into the water within the steel cage, the crew using bait to draw the sharks closer ... it was all standard procedure and, frankly, quite exciting because some of the sharks were huge and sometimes approached within inches of where we were standing on the cage platform that had been lowered into the water.

'We'd been there about 20 minutes when we heard someone say that a very large great white was circling the boat. The next moment we were told to put our heads underwater and I saw this huge graceful beast glide in, straight towards where I was positioned at the end of the group.

'It was quite something to see, because its approach was deliberate and then, with inches to spare, it halted briefly in the water and viewed me with one of its enormous round black eyes that was almost the size of a saucer. I was totally perplexed, even though there were steel bars between us ... I simply couldn't describe the thrill of the

Caroline Castell-Elster, the author's diving partner, was hosted on board Mike Rutzen's boat and spent so much time in the cold water – it was winter – that despite her own wetsuit – she almost ended up with hypothermia. (Photo: Author)

moment afterwards ... something I'll never forget,' Caroline recalls with a chuckle.

Guests who do the dive are all briefed beforehand to listen carefully to the crew member who gives the orders. They stand on the floor of the cage, usually with their heads out of the water and when sharks approach, they are told to quickly duck down and observe the sharks as they approach. That can happen perhaps a dozen times during a session and each time those involved will talk about what an exhilarating experience this is.

Everybody is warned beforehand that the predators are likely to sometimes come within touching distance of where they are perched and that they are not to stick their hands through the bars and try to stroke the beasts. But still, some people do and there have been accidents: nothing serious (yet), but occasional scratches and scrapes from their extremely sharp teeth.

As André Hartman will tell you, some of these sharks might weigh a ton, but they are still able to turn their entire length in the blink of an eye ... one of the reasons why he almost lost a hand in an over-familiar moment of shark handling. It is no different with those watching passing great whites from the protective comfort of a shark cage, he states.

These days, there are thousands of tourists who come to view great white sharks from Mike Rutzen's boat and those of other operators working out of Kleinbaai, a small village a short drive away from Gansbaai, once the shark fishing capital of these southern oceans. These people all pay good money to get on board the boats that take them on the short hop towards the feeding grounds of Dyer Island that are dominated by great white sharks.

A recent article in the *New York Times* described it as follows: 'We pushed off from the dock at 8:30 on an overcast morning and skimmed across the bay for 20 minutes to the marine reserve. Crew members laid anchor, winched the cage into place on the starboard side of the boat, baited lines, and scattered chum — minced fish and shark liver, blended with brine. Then we waited. A shark approaching the boat on the surface elicited the cry of "Shark!"'

As the American journalist described it, he scanned the nearby water and, with a thrill, 'glimpsed an image straight out of the 2003 film *Open Water*: a curved dorsal fin knifing with great speed toward the boat.'

A large great white shark lunges at a chunk of fish used to draw one of these predators alongside. When that happens, guests already standing behind a steel barricade, up to their necks in the water, lower their heads into the sea and view the shark's antics as it passes. Great whites are sometimes only inches away from the viewers. (Photo: Author)

Mike Rutzen

'All 16 of us rushed to starboard and, as the vessel tilted dangerously, we watched the first great white shark of the morning swim toward the baitlines just below the surface. He was a modest six feet long, and though the water was murky, I could clearly make out his torpedo-like shape, beady eyes, and conical snout. The creature lunged for the bait, missed it, and then dived out of sight.'

Having opted for the cage, the reporter together with half a dozen others – all having changed into wetsuits and with masks and snorkels in hand – lowered themselves into a narrow enclosure fixed to the side of the boat. It was a long wait before another shark appeared alongside, which was when he and the others were told to submerge.

Though that experience didn't produce satisfactory results, sharks usually spend lengthy periods alongside the boat while large pieces of fish are dragged across the surface of the sea to draw them closer.

As the scribe explained, 'Then, amazingly, the great white lurched out of the ocean, eyes bulging, jaws snapping, teeth flashing. A ten-foot-long grey mass of muscle and cartilage, he twisted left, then right, in furious pursuit of the bloody chunk of yellowfin tunny. The water frothed around him as he lunged for the bait, and as he slammed his massive body back down into the ocean, cold spray washed over all of us. Some of the passengers screamed in excitement. It reminded me of the climactic scene in *Jaws* when the great man-eater hurls itself out of the Atlantic and devours Robert Shaw on the deck of his boat.

Mike Rutzen

'A few seconds later, the monster was gone, and shortly after that, we started our return to shore. We had glimpsed only three sharks in the course of the morning – about par for the month of February. But those few thrilling seconds had made the entire three-hour wait aboard the *Shark Team* worthwhile.'

Hermanus diver Brian McFarlane – with whom I first dived on the wreck of HMS *Birkenhead*, the British troop carrier that sank near Gansbaai after hitting a rock more than a century and a half ago – has also had his share of experiences off Dyer Island. Brian admits to plenty of brushes with sharks, but as he told me, 'certainly the most terrifying was off that island while spearfishing with my cousin, Richard Kritzinger, for yellowtail ...

'I'd shot a couple and threaded them on a short fish stringer attached to my weight belt. Their fins were continually pricking me so I asked Richard to bring my floatline from the boat. Then I bagged another beauty and was busy sticking it onto the line with the others when I felt the hairs on my neck stand up. I turned and looked straight into the huge dark eyes of a great white.

'When you come face to face with one of these monsters underwater – and it's about four metres long – you can't help sensing that your chances are as good as going into the bush with a *knobkerrie* to slaughter an elephant. It felt like an express train as this great predator brushed right past me. If it had turned with just one little flick of his tail and opened his mouth, I probably wouldn't be here today. I called for the boat and just when I thought I was safe, I saw the shark coming straight at me out of the gloom.

'It was rushing at me with its massive head shaking from side to side and bottom jaw open, typical attack mode. I remember thinking that its teeth looked like steak knives and I was petrified.

'Just when the creature was within touching distance and I thought he was going to get me, I suddenly realised that it wasn't me it was after, but rather the fish I still held in my hand. I immediately released the stringer and in a flash the shark gulped down more than 20 kilos of fish. Looking back, I couldn't help thinking how it all seemed to take place in slow motion.

'By then I'd had enough, so I didn't waste any more time in the water. I hauled myself over the transom and as I sat there it returned once more, probably as a farewell gesture and brushed hard against the hull of the boat. Then it was gone, probably annoyed that it'd missed the main course which should have been me.

'That day was the only time in my life when I felt that death had brushed my shoulder.'

Brian and I dived several times on the wreck of the HMS *Birkenhead*, the paddle-steamer that went down with huge loss of life in the adjacent bay to Kleinbaai.

British soldiers on board had to make their own way ashore, more than a kilometre away. In-between, there were more sharks than anybody could have imagined. Very few of these poor souls survived and some of their graves are still to be seen in surrounding villages.

The sharks are still around, though when you dive the wreck these days, you don't always see them. Most are bronze whalers, two metres long or more, as well as the occasional great white or two.

On our dives, we saw lots of 'bronzies'. They were always around when we surfaced, usually with local fisherman having chummed the water to attract the sharks they catch for a living. All that blood in the sea and these predators having their skulls smashed with the Cape equivalent of a baseball bat, invariably added a tense moment or six to the event.

On a dive with Brian – we set out early one summer morning in his boat from Hermanus to film the wreck – we saw plenty of these beasts. We'd surface with half a dozen bronze whalers around us – some dead and the sea discoloured by their blood, others very much alive. We'd complain to those catching the sharks – as we did each time – but it made no difference. This was how Gansbaai shark fishermen earned their money and they ignored our entreaties. But such things didn't prevent us from diving there either.

Getting down to where the remains of the old wreck lay, was always different. It could also be eerie, with the mind sometimes playing tricks. It was difficult not to imagine the drama that unfolded all that time ago: so many people perishing in seas that were flat and calm. That disaster underscored the danger of a blinder which still sometimes reaches up to within a metre of the surface. In fact, at low tide the rock is visible in the swell.

Our dive boats were almost sucked onto it a few times and one of the first lessons we learned was that you had to look sharp when you operated nearby.

The wreck had other qualities, some sinister. I was on the bottom, totally alone once for about ten minutes – my cameraman having to return to the surface to fetch something – and I can recall that I felt utterly vulnerable. The visibility was perfect once we got through the soup on the surface and there were very few fish and no sharks to be seen. But you somehow sensed something different down there, which is why I never went solo there again. Nor should I have been left at that depth on my own in the first place.

Mike Rutzen, left, with American journalist, author, and television personality Anderson Cooper.

Fishermen at Miller's Point near Simon's Town with a dead sevengill shark. Hopefully this kind of activity will be banned. (Photo: courtesy of Mike Rutzen)

As a diver, it was always distant Dyer Island that intrigued me the most, that black spot on the horizon where the sea churned and the mist would sometimes hang for days. From afar you couldn't miss the noise, the seals' discordant, cacophonic chorus, nor, if the wind turned, the stench, which can sometimes almost curl hair.

Even then, there was a mystique about the place that was inexplicable, in part because the setting was so austere. 'Desolate', one writer described it, and I suppose he was right, because of the early seafarers who came to grief on those rocks and who must have had a hard time of it. Because of sharks, they couldn't simply swim ashore, so they had to wait to be rescued.

I also made a film there a while later, taking our boat between a narrow channel that separates the two rocky outcrops that comprise the tiny archipelago. If you navigate between the islands from east to west, you emerge a few hundred metres downstream between two relatively shallow reefs. Judging by the huge amount of detritus that can be seen cluttering the shore — complete with wooden spars from sailing ships of long ago – it testifies to more maritime disasters than anybody cares to talk about. Who knows what ultimately happened to some of these crews? How many drowned, or got taken by predators? Most times word of the loss wouldn't reach Cape Town until months later.

The island was originally named after a Nantucket black fellow who called himself Sampson Dyer and who worked on the coffin-shaped larger island during the heyday of the American sealers. He'd been stranded there after his ship wrecked in the early nineteenth century. Dyer collected guano to survive, which he sold to local farmers, and, it is said, he eventually settled on the nearby mainland, married a local woman and started a family. His first name, Sampson, is found in some local communities.

Today, Dyer Island is home to about 60,000 Cape fur seals. They are at the top of the menu of hundreds of migrating great white sharks that swarm through the very appropriately named Shark Alley during winter. Though, less plentiful the rest of the year, there is never a time when these beautiful creatures completely disappear.

Good times for viewing great white sharks in the South Cape are from mid-April to mid-September each year. False Bay's Seal Island is a hub of shark-vs-seal-predation. The websites tell us that April and September/October are interim months, with less activity round the boats, but still plenty of hunting seen.

Because these are mainly winter months, the water can be cold, generally 15–20 degrees Celsius (or 59–68 degrees Fahrenheit). The average waiting time for sharks to appear after chumming has started, ranges from ten minutes to about an hour. While it is rare that no sharks arrive, it happens, as it did in the autumn of 2011, with Bruce Gonneau and his family: Bruce is responsible for the layout of this book, so he knew all about great whites and he and his family were eager for the experience. He'd booked beforehand and got his money back because of the 'no-show'.

That said, the Western Cape is regarded as one of the top shark diving destinations in the world and shown as such on the 'Ten Best Places to Dive with Sharks', broadcast by Australian TV channel TV9. The hour-long programme rated the great white shark diving at Gansbaai as the best shark-diving experience in the world. The copper sharks that accompany the sardine run along KwaZulu-Natal's coast were listed third, the shark-diving experiences rated according to how close one gets, the size of the sharks and the 'adrenaline factor'.

In 1994, Mike Rutzen, our host and friend, started working with great white sharks as a skipper on a shark cage diving boat. Four years later he started free-diving with them and is now known

Sijmon de Waal

internationally as one of the few people who is able to interact with these magnificent predators outside the safety of the cage. His activities have caught the imagination of the world and, as a consequence, Mike is regarded as one of the most knowledgeable people in the field today. He has also produced numerous documentaries on free-diving with great white sharks for *Discovery Channel*, *National Geographic*, *Animal Planet*, CNN, CBS, BBC, French TV and Italian TV.

Among the notable celebrities who have dived great whites have been individuals like Brad Pitt, Matt Damon and Prince Harry, while His Majesty King Abdullah II of Jordan has honoured Mike with his presence several times, in large part because he is an outstanding conservationist and passionate about the sharks. Brad Pitt has also gone down three times, Matt Damon twice and other personalities include CNN's Anderson Cooper, Halle Berry and Leonardo di Caprio.

The company also runs what it calls a 'PADI Great White Shark Distinctive Specialty Course' over a three-day period and costs £1,000. The program includes two great white shark cage dives, a scuba-oriented kelp dive with cat sharks (at another location), all meals, transport to and from Cape Town, as well as four-star accommodation. At the end of it, each participant is awarded a PADI certification card. A notable feature of the course is a pair of evening lectures by Mike Rutzen himself: these include the biology, ecology, behaviour and conservation of the great white shark.

There are several other operations based at Kleinbaai harbour, all offering the opportunity to view great white sharks in their natural environment. Costs differ from about R1,000 and up, depending on the season.

In alphabetical order, these include: André Hartman; Great White Shark Tours; Shark Diving Unlimited; Shark Lady; White Shark Adventures; White Shark Ecoventures as well as the White Shark Diving Company.

1   *Sharkman: A Journey Into the World of Shark Hypnosis*: Produced by Off the Fence and Foster Brothers for *Animal Planet*, 2008.

More detail about Mike Rutzen and his operation can be found on his website at www.sharkdivingunlimited.com and e-mail address: info@sharkdivingunlimited.com. Or you may contact the company by ordinary mail at P.O. Box 511, Gansbaai 7220, Republic of South Africa.

Phone bookings can be made at +27 28 384 2787 with a special line for film crews at +27 28 388 0020.

Sijmon de Waal

CHAPTER FIVE

PLAYING TAG WITH GREAT WHITE SHARKS

In recent years, Alison Kock has stepped into the arena on behalf of South Africa's great white sharks numerous times. She leads the research into one of the world's largest concentrations of great whites living on the doorstep of a major city. Alison is a dedicated and passionate shark biologist with a mission to help secure the future of these apex predators through scientific research, awareness and community-based conservation strategies.

Alison's story:

HOOKED ON GREAT WHITES

I will never forget my first great white shark encounter. I had been given the opportunity to join a shark viewing trip to Seal Island in False Bay and what I saw that day, led me to making sharks an integral part of my life. The experience 'kick-started' a most unusual and rewarding journey of discovery: getting to know, and trying to understand, arguably the greatest predator of our time, the great white shark.

On reflection, I can still sense the excitement and anxiousness of the day, not really knowing what to expect. What I certainly never envisioned was a great white shark of a ton or more launching itself completely out of the water right next to the boat as it breached, causing the wave that it created when it crashed down into the water again, to rock the boat from side to side. It was a most remarkable experience.

These days – more than a dozen years later and following hundreds of trips to Seal Island and surrounds, coupled to thousands of hours interacting with great whites – I am still just as thrilled when it happens; in my mind, you simply can never get enough of these magnificent oceanic creatures.

Each day I learn something new and am presented with new challenges and revelations. I simply cannot think of a more rewarding and fulfilling way to live my life. Nor do I want to do anything else.

WHY TAG SHARKS?

The truth is that though we view, study and come to comprehend many things about sharks, these predators are still barely understood. More salient, large sharks are increasingly under threat, not only from overfishing, but also from the destruction of their habitat, loss of their prey and pollution.

In truth, sharks are especially susceptible to all these impacts, especially those like the great whites which tend to spend lengthy periods in coastal waters, adjacent to large urban areas like Cape Town. It is part of my job to better understand how sharks manage to exist in these parts of the ocean so that we can identify and counter some of the potential threats to their populations.

Comprehending how important these areas are to the predators is an essential step in the kind of conservation in which I and others are involved. But in order to do this, we need to be able to monitor their movements, their actions, how they live and feed and all the rest that make the world go round for these ocean hunters.

Modern technology has played an increasingly important role in this work. The advent of acoustic telemetry, for instance, has enabled us to monitor sharks in their undersea domain for extensive periods.

There are other aspects, like the conflict between shark and mankind, which is just one more threat that faces these large predators, and, in particular, great whites. The scientific community working on these problems is well aware that traditional steps taken to mitigate the problem, such as the erection of gill nets, drumlines or the execution of shark culls to reduce shark populations (which, in turn, reduces the risk of an encounter), does have deleterious impacts on our sharks and marine ecosystems.

We must move towards methods of reducing conflict, which are based on a sound understanding of shark presence and risk. Thus, we will have a better understanding of their movements, such as areas and times of high activity, and these efforts can help us avoid these areas in bids to reduce encounters. Similarly, identifying migration routes or 'hotspots' helps us to avoid those areas as well.

And finally, a more selfish reason for tagging sharks has been a quest for a better understanding of these magnificent and rare animals. Studying them up close has allowed me to see that everything I thought I knew about great whites before I started with this work was either false or inaccurate, and that I needed to contribute to a new chapter for sharks, one based on reality and respect and not fear and fiction.

FIRST TAGGING ATTEMPT

The first time I had to tag a shark, my emotions were decidedly mixed. I was obviously excited at the prospect of doing something so adventurous. Also, it was the start of a project towards my Master's degree and I was eager to tag my first shark. At the same time, I was more than a little anxious, and among other fears, I was worried that I might tag the shark in the wrong place and cause it to be injured, or miss completely and lose the expensive tag.

In tagging sharks, it is important that the tag be inserted directly below the dorsal fin: it must be firmly embedded in the shark's thick musculature, where it remains in place for an adequate period. This is a part of the shark's torso where there are no major blood vessels or vital organs, thus little risk of injury to the shark.

To accomplish this task, it's necessary to work with a dedicated and coordinated team of people. But first, we needed to get close to the shark to achieve that objective. The job starts with chumming the water in order to draw the shark closer, but even this task is not always easy because it is the other sharks that choose to come along that makes life difficult. Sometimes there are days spent watching these predators going to work on the seals in the water around us, and when that happens, we are stuck with a handful of what we like to term as 'drive-byes'.

Terminology in this industry is catchy and I soon learnt to use words like 'players', 'drive-byes' and a host of others. Players are sharks that interact with us more than others: they are generally more relaxed, but also motivated to follow the bait or decoy. Indeed, we usually get just about everything we need from the 'players', their size, sex, photo-ID, genetic sample and tagging – together with some marvellous behavioural information.

'Drive-byes' as the expression implies, means that the shark simply swims right past the boat and almost no information is collected from that animal. Having just 'drive-byes' is obviously frustrating, especially when there is pressure to sample and tag enough sharks in order to collect data. Consistent chumming means that we get the best opportunity of attracting a variety of individual sharks to the boat to tag. Usually we use the head or tail of yellowfin or longfin tuna donated by members of the fishing community.

Once the shark starts to show interest in the potential free meal, our job is to estimate its size — to the nearest half-metre at least — sex it, and gather information on any identifiable markings that distinguish it from another predator. Thereafter it's a process of photographic identification of its dorsal fin to include in a photo catalogue of all sharks encountered. All this needs to be done *before* the tagging takes place, otherwise the data we collect is not as useful as it should be.

At this point, the bait handler probably plays the most important role in the process because he or she has to handle this enticement in such a way as not to allow the shark to lock its jaws into it: the same piece of bait might be used many times to draw the predator closer. At the same time, the bait handler has to keep the shark interested enough to keep it at bay and remain in the immediate vicinity of our boat.

What is important is that the shark doesn't get the bait because if it does manage to clamp into the bait, it often ends up thrashing about, which can make accurate tagging almost impossible, and more importantly, we want to limit our impact on the sharks by not providing regular 'free meals'.

There is a tremendous amount of skill in finding the balance of keeping the bait immediately in front of the shark's nose without it sinking its jaws into it, and there are some people that I simply believe have a gift for this. My husband, Morné Hardenberg, a shark diver and underwater cameraman, is one of these people and I believe this gift is one of the reasons I fell in love with him in the first place. People like Morné have the ability to read the shark's subtle behaviour and anticipate its actions, and react accordingly. In a sense, it's a bit like a dance — backwards and forwards, to-and-fro, as well as giving and taking.

In the beginning, I used a sturdy 2.5-metre metal pole in order to insert the tag. As soon as a shark is successfully lured close to the boat, I have to position myself alongside the gunwale and hold the pole ready for probing the creature. Once I am happy the shark is close enough, I bring the pole down sharply to jab the tag into the base of the dorsal fin. But, experience has proved that this technique can be inaccurate: by the time you have brought the pole down in the initial thrust, the shark might have turned in the water and that could mean that your tag could end up in the wrong part of its body. Or you could miss completely if the shark is swimming too deep.

I remember a particular day when a team of us took Mike Rowe, the American television personality from the popular American prime-time series *Dirty Jobs* out with us on a tagging jaunt. We were filming for the series which would show us at work.

Mike was incredibly funny to have on board and it was an entertaining day, even without the sharks coming up close. He just couldn't seem to get over the fact that sharks have a pair of claspers which they use when mating. His jokes were endless on that topic and he kept us in stitches. It's important to be aware that having a sense of humour is essential for being a shark field biologist, because if you can't laugh about broken-down boats, faulty research equipment, poor shark action, inaccurate weather reports and be able to accept that some days you were simply meant to stay in bed, then you should scrap the idea of working in this field.

When the time finally came to tag sharks, the boat we were on was a lot higher out of the water than my research vessel. That basically meant that for me – I was going to tag the first shark — the tagging angle was particularly uncomfortable. Also, I was more nervous than usual because the cameras were rolling and my fellow colleagues, all experienced in this field, were on the boat with me. This was the first time they would see me tagging, and as a young woman researcher, I felt a lot of extra pressure to do it just right.

When the first shark was lured in, I hesitated and aborted the attempt because I wasn't confident enough it was going to be inserted correctly. Luckily it was a 'player' and I had another chance to tag it, and when it swam around again I followed through and tagged it. Unfortunately though, I'd got the tagging position wrong and the tag wasn't placed on the shark as I'd originally intended, it was too far forward, which, though not harmful, was not ideal. That was when I decided that there had to be a better method that would eliminate all that anxiety and inaccuracy with the tagging process.

Enter the modified speargun and it was here that shark specialist Mike Rutzen provided the solution. Having explained my frustration with the unwieldy process of using a lengthy pole to place shark tags in position, Mike suggested that perhaps I should think about a modified speargun for the task. His concept was to use a speargun where the spear was welded to the barrel and only allowed to move about 45 centimetres once fired in order to embed the tag into the tough skin of the target shark.

Together with an engineer, we developed the new device. We adjusted the thick rubber bands and added a rubber stopper to the

Morné Hardenberg

end of the spear, so as to have the least amount of impact on the shark. This way we would achieve just enough force to implant the dart into the shark with the least amount of force. Once the gun was ready, I got hold of some longfin tuna and fired a series of test shots into the fish on my balcony, much to the amusement of my neighbours. It performed perfectly on the chunk of tuna fish, and I was satisfied that it was safe to test on the sharks.

The next day, with Morné in tow, I took the gun to sea and was happy when the first shark was finally lured in. I leaned slightly over the gunwale, placed the tip of the spear about 30 centimetres below the predator's dorsal fin and pulled the trigger. Voila!

The tag was no longer on the spear and the spear was still attached to the gun, but in the meantime, the shark had swum around the boat so I wasn't able to see whether the tag was in or not. Morné said he saw the tag go in, but I wasn't so sure: I needed to see for myself. It seemed like a long time before the shark swam around once more: to my delight the tag was perfectly implanted in the exact place I wanted it and the shark carried on swimming as if nothing had happened.

The speargun concept was a huge success, and since then I have confidently and accurately tagged over 70 great whites. And also successfully tagged sharks with many of my colleagues, and even CNN's Anderson Cooper, Jeff Corwin and the Smithsonian's Paul Rafaelle, each time with a remarkable degree of accuracy and much less stress!

DIVING IN 'SHARK HOTSPOTS'

Once I'd decided that we'd tagged a representative number of sharks around Seal Island, I turned my attention to the listening stations, the devices that collect the impulses sent out by the tags. These are self-contained units that record the unique code identifier, date and time (and in some cases, pressure and temperature as well). In this way I can determine when the shark arrived, when it left (and thus how long it was present), which shark it was and whether other tagged sharks were recorded simultaneously.

These listening stations have to be retrieved every few months in order to download the data, change their batteries and clean bio-fouling, which can affect performance of the equipment. Our method of doing this is to prepare complete devices which can be swapped out with the existing stations, so that we can download the files at home, at the same time ensuring continuous monitoring of the area.

Diving off Seal Island is a logistical nightmare, and strict codes of conduct are enforced to ensure the divers' safety: the place has seen many shark attacks in the past, some fatal. On top of normal risks of diving, such as decompression illness, we obviously had to consider the possibility of shark attack.

Due to the nature of the objectives of the project, the listening stations needed to be placed in areas with the highest great white activity, which included the prime hunting grounds around the island.

In an area where great whites are world-renowned for their ambush-attack strategies on seals, you can perhaps forgive divers for feeling a little apprehensive when they enter the waters here. As a result, we always ensure that we drop them on the exact mark we recorded when we originally placed the stations, or certainly as close as possible. By doing so, we limit the need for the divers to have to search for the instruments.

Furthermore, we deploy a safety cage for the divers to use when descending or ascending in the water, but the cage also needs to be dropped on the exact location and has to be relatively stable, so calm waters, little wind and current are what's needed. If these conditions are not fulfilled, a cage ends up becoming a liability rather than a safety measure.

Once everything is in position, which can take up to an hour, the divers quickly get into the water and descend. Their most vulnerable moments are on the surface, alongside the boat and on the way down, so getting to the bottom as quickly as possible is essential. Once at the bottom, the divers usually locate the listening station and then start to loosen the bolts that hold the system in place on its mooring block. Obviously, good visibility makes for a safer and easier dive, but this being Cape waters, visibility is usually less than five metres, with some areas like Strandfontein and Macassar limited to about a single metre. These situations add considerably to the threat factor.

Our primary research divers are experienced and have been assisting our research since 2001. Clearly, their experience makes all the difference as to whether the task is successful or not. Also, there is the all-important issue of money when tackling this task.

On the one hand, while every station costs roughly US$1,200, the data that these devices provide the scientific community over a single season is priceless. So we don't want to lose any of them either! Ensuring that the old instrument is recovered intact and the new one correctly and securely deployed is essential to the ultimate success of the project, not to mention determining how much sleep I manage to get in the interim.

Every minute the divers are down below, I hold my breath, ready to react to any scenario such as a diver having problems with equipment, ascending too quickly, or being buzzed by a shark. Typically, if they remain underwater longer than ten minutes, I start to sweat a little. Everyone on the boat has an important role to play, but there is none as vital as lifting the cage once the divers are ready to ascend from the bottom of the sea. Going up at a rate greater than 18 metres-a-minute can be dangerous and if a diver has a problem such as a reverse block, then we need to know about it immediately and stop the process.

It goes without saying that all this means exceptional communications between divers and the boat. Also, lifting the cage is no easy matter, but fortunately we have willing and strong volunteers who do most of the work. Handled in this manner, each of these station retrievals at Seal Island takes roughly two hours.

Of course, the most exciting event for me is to rush home afterwards, hook up the station to my laptop and download the data. The time it takes to download is correlated to how much data has been stored on the station. If it takes a while, then we are likely to have some great data available, which makes everything worthwhile.

The amusing part is to see how many tagged sharks were detected while the divers were down, and then, tongue-in-cheek, to let them know about it.

A SPECIAL SHARK OF SEAL ISLAND

During the course of our work, we have had some spectacular days at Seal Island. There have also been quite a few treasured memories.

Working so closely with a relatively small population of large predators, means that we get to know some of these sharks really well. One particular shark called Nutcase is an animal that comes with quite a history. On a beautiful day in 2004, we were lying at anchor off the south end of the island and had recorded 14 different sharks at the boat. Shortly after midday, this activity slowed down to almost no sightings at all. However, after about 20 minutes, we spotted a dorsal fin slice the water nearby and I immediately noticed a distinctive natural tip on the shark's dorsal fin which made it stand out from the rest.

This shark was a male of 3.2 metres, and while great whites are usually cautious when they first approach our boat — typically they will make a few passes around and under the hull until they have built up their confidence to get closer – this shark was different. Smaller sharks sometimes deviate from this cautious behavioural pattern when large sharks are about: they often rush the boat and the bait in an effort to get in and out, so to speak, before the larger sharks return.

The behaviour of Nutcase — while bold and erratic at times — was mixed. Sometimes the shark was quite relaxed. Then it would suddenly become quite animated and determined to sink its jaws into our motors, with great vigour at every opportunity. Nutcase once even stole our chum bag that had been hanging over the side of the boat between the motors. At other times it would swim away and mouth and bump floating kelp and even stick his head out of the water and take a good look at us.

That first time Nutcase spent over an hour circling our boat, apparently quite intent on making us believe that he was mad as a hatter, thus earning himself his rather appropriate nickname.

Nutcase ended up becoming one of our favourite sharks and this animal made a huge impression on me because of the amount of time we were able to interact and gather insight into his habits. However, he would rarely approach the boat if there were other sharks about. Instead, he would wait for a gap and then return to us again.

We recorded Nutcase on five different occasions that year and tagged him to monitor his movement patterns, but later in the season, he started spending less and less time with us. In our scientific publication on the effects that chumming has on shark behaviour (Laroche et al, 2007) you can read all about this under the heading 'SHARK 31'.

In 2005, Nutcase returned to Seal Island and though he had lost his tag by then, we immediately recognised his very distinctive dorsal fin and characteristic behaviour.

Naming sharks might seem like an unscientific approach to some, but it allows for a quick and easy reference to identify animals that are fairly well known to those of us working with them. It's far easier to recall a name after an extended period of time than a catalogue number, especially when you are dealing with many different creatures at once.

We recorded him on four different days that year. In 2006 and 2007 he returned again, but only approached our boat once each of those years. In 2008 we had no sightings of Nutcase at all, which was disappointing because these re-sightings yield valuable information, such as how much the sharks have grown, whether they sport any new scars, and other info that can be used to estimate population size. On a personal level they bring you closer to the animals that you spend your entire life trying to learn more about.

We are not the only boat recording sharks at Seal Island. There are also shark cage diving boats that operate in the area. We have shared experiences with them and it didn't take long to discover that they had also experienced interactions with Nutcase. Interestingly, they never saw him in 2008 either.

When you don't have the usual encounters with specific sharks for an entire season, you begin to wonder about what may have happened. There are numerous possibilities: these can range from it simply being part of a natural migration cycle we have yet to understand, or it could be something more sinister, like being caught or snagged in fishing gear. Due to some of the difficulties we are faced with in studying great whites, we never know what has happened when a specific shark totally disappears off the radar.

Luckily, this was not the case with Nutcase. In late 2011, we had been having an excellent day with about a dozen different sharks recorded. Keeping tabs on the individuals needed a lot of concentration and I was totally focused on the task at hand. Then, quite suddenly, we had a large foam slick floating past our boat from the island, which looked almost like a white carpet on the water surface. One of the interns brought my attention to the fact that there was a shark investigating some kelp in the foam and when I looked up I immediately recognised Nutcase by his distinct dorsal fin.

I was delighted and screamed with joy so loudly that those on board the other boats in the vicinity all looked our way.

Nutcase slowly approached and I could see that he had grown significantly to almost four metres and was now a mature male. He spent a few minutes with us displaying some of his characteristic behaviour, such as spy-hopping that we knew him for, but he was also more relaxed and confident. One of the shark operators called us to say he had also seen Nutcase and later he shared with us some incredible photos of him hunting seals.

It was a reward to encounter Nutcase once more and I can only hope that we get more special visits from him in the years ahead.

TAGGING SHARKS CLOSE TO SHORE

I started this project at Seal Island and my focus was to identify the hunting strategies and residency of sharks around the colony. However, two years in, Cape Town experienced a number of shark incidents in inshore areas along the coast, all of them involving great whites and some fatal. This was a turning point in the research and my focus shifted to gaining a better understanding of their presence close inshore, particularly at some of the more popular beaches, like Fish Hoek, south of Muizenburg.

As a result, we deployed 35 listening stations along the coast in False Bay alone, stretching from Cape Hangklip to Cape Point, a distance of more than 150 kilometres. This meant that I had to considerably expand my own activities in order to tag sharks closer to the mainland and to ensure that I covered a sample representation of the great white shark population that sometimes moved inshore.

This raised a few problems, one of which meant that when interacting with sharks close to the coast, there always had to be a concerted amount of logistics and coordination with adjacent beaches. Of course, there was a lot of media and public interest and that obviously meant that we were in the public eye a lot more than we liked. Taken together, all these factors added significantly to the pressure.

The first time I tried to tag inshore was in September 2006. I was anchored off Seal Island when I received a call from one of the shark spotters at Fish Hoek. That was from Sabu – who has since passed away – and he told me that a very large shark had just entered the bay. I immediately called the authorities to get permission to attempt to tag that shark and raced over there.

When I arrived off Fish Hoek, the beach and adjacent areas had been closed because of the shark. One of my crew was on the radio to the shark spotter who then guided us towards where the shark was swimming.

Fish Hoek that day was like a tropical pool, beautifully flat and clear, which would have been almost perfect conditions for the swimming community (who were now all out of the water and waiting on the beach). It was also ideal for tagging.

When I first saw the shark, my heart took a leap. Not only was this shark swimming about 50 metres from shore and in less than two metres of water with its tail sweeping across the sand, it was also the biggest great white I had seen in more than a year. I estimated it conservatively at about five metres.

Sabu, the shark spotter sitting on a rocky ledge high above the beach, was excited too. He was telling us over the radio to get closer and try to tag it.

Contrary to what people might think about these great predators, they are extremely cautious, and at that point I didn't want to frighten her. I assumed it was a she, given her large size and the fact that female great whites are larger than their male counterparts. I followed her at a distance for a few minutes to gauge her behaviour.

At this point, the shark was swimming quite fast, at about four kilometres an hour, or roughly the speed of a brisk walk on land. Also, the shark was moving with purpose. The reality of the situation was that you cannot simply approach a shark and invade its space, expecting it to accept you: you need to try and anticipate what it wants to do and work along with it.

Since this was my first time that I had tried to tag a shark without being anchored, nor being able to chum, the shark was clearly on an as-yet-undetermined mission of its own. I quickly devised a plan, and that was to try to position our boat directly in front of her path and then throw a tuna bait in her way. I was hoping that the chunk of fish would attract her attention so she would interact with us. We did this, hurled out a piece of bait, but she completely ignored it. So we tried it several more times, but she either just swam past, or dived underneath the boat to avoid us. It was clear that this shark was not interested in interacting with us.

By now, I was feeling the pressure. I desperately wanted to tag this massive female. Also, I was getting frustrated. At that point I decided on a more direct approach: the shark was now headed out the bay into deeper waters and I was going to miss my opportunity. As a last resort I asked my crew member to drive right up to her and see whether I could lean over and possibly jab a tag into her. This approach resulted in this massive animal charging us and sinking its jaws into the side of the boat. It was pretty clear that she wished to be left alone, which is what I then did as I watched her dive deep and out of sight.

Shark spotter Sabu — still on his perch on the high ground above us — was furious. I had lost an excellent opportunity to tag a shark and he made his views quite clear over the radio, telling me how I should have done it.

Deflated, I started to call some of my colleagues to let them know that I hadn't t been successful and that the shark had left the bay. But then, barely 15 minutes later, I thought Sabu was getting his revenge with a practical joke by radioing through that another large shark had just entered the bay, this time on the Sunny Cove side of Fish Hoek.

We raced over and yes, there was another shark, perhaps a metre smaller than the first female, but displaying the exact same behaviour: she swam at the same speed, along a roughly similar shoreline contour.

As I approached, I saw that this shark already carried one of my tags, which excited me because I was aware that it had come from Seal Island and here, quite unexpectedly, she was in waters off Fish Hoek. All I could think about was the fact that she was swimming past all four of my Fish Hoek listening stations and I would be able to identify her tag code.

This time I kept alongside the shark at a distance of about 15 metres and simply recorded her behaviour. Like the earlier visitor,

Sijmon de Waal

she swam parallel to the shore, also in about two metres of water and kept going until she was out of the bay.

To my utter disbelief, and starting to wonder whether I hadn't spent too much time in the sun, it was less than ten minutes later when Sabu came on the radio again to say that a third shark had entered the bay on the Sunny Cove side. What an extraordinary day we were experiencing. This shark was the smallest of them all and swam a little slower than the others along a somewhat different pathway. The moment we were a few metres away, she started to interact with us by circling us widely. Again, I tried a few different techniques to get her to follow the bait and come close enough to be tagged, but none of my ploys worked. I did get one attempt to tag her, but the speargun jammed and it is possible that the noise frightened her away. We did however, manage to get her dorsal fin photographed, so we could identify her again.

After all this excitement, it was a day I will never forget. The three shark sightings also created quite a stir, with one of Cape Town's newspapers running a headline which read: 'Great White Eludes False Bay Tag Team'.

However, as far as I was concerned, I was delighted: one of the sharks had already been tagged and seeing them like that, swimming naturally in that bay and in a completely different habitat that I was used to encountering sharks in, was something of a highlight. Since then I have successfully tagged a dozen sharks in this manner, close to shore and every one of them females.

GREAT WHITE LEADS US AROUND THE CLOCK

Another method of monitoring the very fine-scale movements of sharks is to tag them with a continuous 'transmitter'. This module sends out a constant pulse, which we are then able to follow by using a hydrophone mounted on the boat.

Following animals at sea is challenging and labour-intensive. Also, we need to stay with them, allowing for as little disturbance on their natural behaviour as possible so that the results we collect are not compromised by our presence, which we now know, can alter the shark's normal patterns ...

On 9 December 2011 we tagged a female great white of 3.2 metres with a continuous tag off Strandfontein, some distance further down the False Bay coast. We dubbed her 'Deepblue', and the tag allowed us to follow her every move. While we have tracked sharks in the bay before, this was to be the first successful overnight track, enhanced by the fact that conditions were calm and flat and blessed with clear skies.

We were lucky this fine summer's day and it only took us 30 minutes to spot our first great white patrolling an area parallel to the coast. It was a small shark of approximately two metres. Unfortunately, as we approached, it decided that it wanted nothing to do with us and swam away. Luckily, only minutes later, another great white – larger and more confident than the one before – was spotted. We were able to get quite close to her because she was curious and approached the boat. I promptly tagged her.

Immediately we set up the system known as the VR100 – research equipment that is deployed from the boat and used to monitor the tag. The system was a scholarship provided to me by Vemco Ltd and has proved invaluable for researching fine-scale movements of large sharks. As we switched the device on, the expected ping...ping...ping came through loud and clear. That told us that the shark was close by, but on the move. While tracking her, we observed some rather interesting behaviour: she met up and followed a second shark of similar size for a short while when they were both at the surface.

There was no obvious sign of aggression from either of the predators, but we were able to observe that both sharks were wary of each other. Consequently, the interaction was over in less than a minute. There were two more close interactions with other sharks, also lasting only seconds. Our original target shark spent a fair amount of time swimming about on the surface where we could view her quite clearly. She was in no rush, sometimes moving along at about two kilometres an hour, and, at best, increasing her speed to about twice that. We were treated to a breathtaking sunset at the start of our first overnight track.

There's something a little omnipresent about being in a small boat at night when you're following a large shark around, hearing, but not being able to see the waves crashing onto the shore. But these night-time activities also provided me with some of my best memories. Once all the sun's rays had been sucked in by the dark sky, I realised that there was bioluminescent plankton in the water; microscopic sea creatures that can produce light through chemical reactions taking place within their bodies, called bioluminescence. The biolu-minescence results from a light-producing chemical reaction called chemiluminescence.

So, when disturbed as the boat moves through the water, plankton lights up the water with a faint yellow glow. On a previous night research trip at Seal Island I had seen sharks and seals glide through bioluminescent plankton and it was a magical sight. The disturbances created by the sharks and seals resulted in faint yellow glows outlining their silhouettes, giving the impression that they had a halo around them.

Seeing the same conditions on this night, with each hour that passed my anticipation grew, hoping to once again to be able to observe such a sight. However, what actually happened was more than I expected. It was just after midnight and we'd been tracking the shark for about 14 hours, and all she was doing was swimming very slowly in a roughly circular pattern between the breakers to about two kilometres from shore while staying in a small area.

The team was tired, and hungry (by now we'd already eaten all the food) and longing to be in our beds, when all of us witnessed an explosion of light just behind the boat. A fountain of yellow light followed by a loud splash burned an image in my mind. 'Deepblue' – or another shark – had just breached out of the water, creating something impossible to forget. This caused a stir and invigorated us for another few hours, hoping she might breach again. Finally fatigue, cold and hunger took over and we headed back to base in Simon's Town.

WHAT HAVE WE LEARNT?

Our understanding of great white shark behaviour is slowly reveal-ing a much more complex animal than any of us originally anticipated, and, as a consequence, I believe the tide is slowly turning for sharks. Research provides a solid fountain on which to build conservation campaigns, but it's of limited use if not applied to education, awareness and policy changes. Most importantly, whole communities need to get involved.

The tagging process did have its share of revelations, the first being that there is a year-round presence of great whites in False Bay, coupled to a consistent seasonal migration of sharks, which aggregate around the seal colony from April to September (the winter months). That contrasts with aggregations close to shore over the summer period. The switch occurs over a relatively short period and starts when the Cape's southeasterly winds become more frequent.

Tagging has also revealed that great whites spend many months close to popular beaches and that these areas are just as important as seal colonies are for some sharks. Most likely, diversions from the island (and seal) environments are used for feeding on fish or other shark prey, and also for resting. Generally speaking, the average great white has a fairly predictable behaviour and aggregations are customarily timed with an abundance of prey.

Residency times vary between sharks, with some animals that are highly site specific (they re-use the same area over and over again). We also know that sharks can swim between False Bay and Gansbaai which lies about 150 kilometres to the east, within less than two days and while some do so frequently, others were not recorded doing anything like this.

The tagging process has allowed us to compare what we were observing on the surface to the sharks' presence when out of sight, and also taught us to be cautious when interpreting observed behaviour on its own. One of the very intriguing things we discovered was that just because the sharks were not attracted to our boat – or were not seen hunting seals on the surface – that did not mean they were not there. Frequently our results showed that tagged sharks were in the immediate vicinity of where we were, but were simply not interested in showing themselves.

Contrary to our expectations that chumming and baiting at Seal Island would have a significant impact on shark behaviour, we found that tagged sharks sometimes stopped responding to us over the season. It is my view that the reason might lie with the fact that current operations at Seal Island offer no regular or substantial food 'reward' at the boats to make it worth their while. The bottom line here is that sharks are interested in filling their bellies, not simply hanging about on the surface of the sea with little reward.

Overall, the sharks at Seal Island actually have very little contact time with boats. However, continuous monitoring is essential to make sure it stays this way.

So far, we have tracked specific sharks over one or two seasons, a snapshot in the life of an animal that can live to over 45 years. We need to understand how they use different areas along our coast over different stages of their life cycles. We have to ascertain too, whether their actions might differ between sexes, something which we strongly suspect. This will have implications for exactly how we apply managerial strategies.

There are many things we still need to learn about great whites, which puts me in high spirits: it means that I can spend the rest of my life getting to know these magnificent ocean predators and spend my time in a world that's still their domain.

Long may it stay that way ...

Sean Botha

CHAPTER SIX

SHARK ATTACKS HAPPEN:
THE REALITY

The headline in a BBC news report on 28 September 2011 said it all: Swimmer loses legs after shark attack in South Africa. The 43-year-old man, 'who ignored warnings to stay out of the water at a beach near Cape Town', entered the ocean 90 minutes after the shark had been spotted and the beach closed.

Michael Cohen, 43, a British resident of Cape Town was attacked by a 3.5-metre great white shark that had apparently been spotted in the vicinity of Cape Town's Fish Hoek beach.

No stranger to the area and a regular in False Bay waters, Cohen might not have been aware that a warning siren had already been sounded. At the time of the attack he was perhaps one of maybe two or three people still in the water, though another report maintains that he was in relatively shallow water when the attack happened.

People on higher ground above Fish Hoek beach – there are houses that stretch some distance up the hill to the immediate south – spotted the shark circling Cohen, but it struck before anybody could warn him. According to the National Sea Rescue Institute his right leg was completely severed and his left leg removed below the knee.

Of significance was the fact that several bystanders witnessed the attack, including a Mr Drysdale, 61, and a Mr Till, 66 – both of whom had been working as 'spiritual guidance volunteers' at Cape Town's maximum security prison and were driving home when they spotted the lone swimmer and then the shark.

Without hesitation the two good Samaritans kicked off their shoes, plunged into the water and waded through the surf to reach the injured swimmer. Drysdale had called the emergency services before going into the sea himself. The two men fashioned an improvised tourniquet from a wetsuit and managed to stem the blood-flow, which, surgeons said afterwards, saved Cohen's life. Shortly afterwards, he was airlifted to hospital.

Less than two weeks later, on 10 October 2011, another great white shark attack made the news, this time in Australia. It was the third fatal attack in a two-month timespan, this one involving George Wainwright, a 32-year-old Texan who had been scubadiving when a witness on the dive boat saw 'a large amount of bubbles'. Immediately afterwards a body came to the surface with fatal injuries. A three-metre great white shark was later spotted in the area.

Rottnest Island, where the attack took place, is roughly 18 kilometres from a beach in Perth where another man was believed to have been killed by a great white. In the earlier attack, the victim failed to return from a swim at the popular Cottesloe Beach. Police divers later found the swimming trunks of 64-year-old Bryn Martin on the sea floor, and reported that the damage was consistent with a shark attack.

The deaths follow that of a bodyboard surfer from a shark attack near Dunsborough, also in Western Australia, a month before.

Only months later, in April 2012, 20-year-old Springbok body-boarder David Lilienfeld was taken by a massive great white shark in False Bay. The incident took place while the victim was in the water with his brother Gustav at a remote location between Gordon's Bay and Rooi Els. His father, Dr Dirk Lilienfeld raced down from the road and was present when his son's body was taken from the sea. It became clear from witness statements afterwards, that this promising young athlete was attacked twice 'with great purpose' by the predator – reckoned to have been about five metres long – which then disappeared.

Earlier, on 28 June 2011 and also in South Africa, the diving fraternity had already been rocked by bad news. An American diver working with a research team studying the annual sardine run up the east coast of southern Africa had been attacked by a three-metre shark just off Aliwal Shoal, a 40-minute drive south of Durban, Africa's biggest port.

It took time for details to emerge. Very little information was given to start with, some of it contradictory. For a start, the annual sardine migration had not yet arrived in that area in any number, even though an initial statement by the firm involved following the attack did mention sardines.

It later transpired that indeed, there were sardines but these were of the frozen variety and the boat involved had been chumming sharks at the time in order to attract tiger and other sharks so they could be viewed by paying guests. Another report stated that 'at the time of the shark bite on 22-year-old Paolo Stanchi, there were a few sardines around, but nothing spectacular. The shark was a dusky, not a bronze whaler as some had said, and was about 3.5 metres long. Apparently, it circled several times, disappeared, reappeared and then targeted the young fellow. It was a baited shark dive ...

Those who take part in this kind of underwater activity are aware that Aliwal Shoal is a popular recreational and research dive site, and known for diver interaction with raggedtooth sharks, which are generally believed to be passive, at least in warmer Indian Ocean waters. They can be quite aggressive further down the coast towards the Southern Cape.

What is sad is that young Stanchi lost a leg and he had an arm severely mauled in this attack, which was apparently quite vicious. It was significant to this writer personally, because only months before, he had taken his partner Caroline diving in the same area. She had only a short while before completed her dive qualifications – BSAC in the UK and PADI with Marc Bernardis's Shoal Divers in Umkomaas.

Meantime they had both spent time in the water with a couple of dozen blacktips and nothing untoward happened. Nor was it expected to, which underscores this event as being unusual. It was the first serious shark attack off Aliwal in many years.

The attack on young Stanchi was one of a spate of more than half a dozen shark onslaughts during the summer of 2011, an unfortunate trend that does not bode well for the future. A newspaper headline in *The Independent* that August said it all: 'Worldwide Shark Attack Hotspots Flare Up'.

It mentioned five attacks in that and preceding weeks alone: two at the same holiday beach in the Seychelles that killed British vacationers. Scientific analysis since has indicated that one or more tiger sharks in the four-metre range were involved on a British honeymooner. Two attacks in Russian waters followed and then there was a 27-year-old tourist attacked off Puerto Rico. At about the same time, a South African holidaymaker was killed by a large shark at Plettenberg Bay in the Cape and an Australian surfboarder bitten in half by a large great white shark a short time later.

A more serious – and widely publicised – shark attack took

place off Fish Hoek in the Cape Peninsula more than a year ago, in January 2010. This involved Lloyd Skinner, a Zimbabwe national and former graduate of the University of Cape Town. Skinner, aged 37, was standing chest-deep in the water only 20 metres from the beach when he was taken. The shark was described by people who were there as a huge predator, at least four metres long.

Phyllis McCartain from Arundel in Sussex said that she saw the shark come back twice, adding that 'it had the man's body in its mouth, and his arm was in the air. Then the sea was full of blood.' Her friend, Denis Lundon, added that the swimmer was thrust chest-high out of the water.

The truth is, shark attacks happen. It is a reality that these events generate sensational interest, out of all proportion to the event. During South Africa's holiday periods at the end of the year and over Easter weekends there might be a thousand deaths in road accidents in the country, but a single shark attack will focus the attention of readers like nothing else.

It is interesting that only days after the attack on Lloyd Skinner in Cape Town, Heidi Blake of Britain's *Daily Telegraph* listed what she called 'The Top Ten Worst Shark Attacks', which, ironically, was carried by another website that displayed a magnificent whale shark across the top of her report. The truth is, whale sharks survive on plankton and have never been known to have threatened humans. Her report reads as follows:

1) USS *INDIANAPOLIS*

More than 900 men were thrown to the mercy of the sharks of the Pacific Ocean when their American warship, the USS *Indianapolis*, was split in two by Japanese torpedoes in July 1945. When rescuers arrived four days later, they found 579 men dead, with many chewed to pieces by circling sharks. Woody James, among just 316 survivors, said later:

'The sharks were around, hundreds of them ... Everything would be quiet and then you'd hear somebody scream and you knew a shark had got him.'

2) SHIRLEY ANN DURDIN

In 1985, Shirley Ann Durdin was diving for scallops in Australia's Peake Bay when she was attacked by a great white shark, said by witnesses to have been 20 feet long. The mighty fish tore the 33-year-old in half in its first strike as her husband

and four children watched in horror from the shore. By the time rescuers arrived, all that remained was her headless torso floating in the water. Within moments, the shark returned and devoured that too.

3) JERSEY SHORE ATTACKS

Four people were killed in a spate of shark attacks along the coast of New Jersey in the United States during the deadly heat wave of 1916. The first victim, 25-year-old Charles Vansant, bled to death after sharks stripped the flesh off his thigh as he went for an early-evening swim. Five days later, Charles Bruder, 27, was killed after a shark tore into his abdomen and severed his legs as he swam off the beach at Spring Lake, 45 miles north of Beach Haven. The final attacks took place six days later on 12 July in Matawan Creek, 30 miles north of Spring Lake. Lester Stillwell, a 12-year-old local boy, was dragged underwater as he splashed in the creek with friends. Stanley Fisher, 24, plunged into the water to search for Stillwell but was himself attacked by the shark and bled to death. The boy's mutilated body was found washed up 150 feet upstream two days later. A fifth man was also attacked later that week, but survived.

4) ROBERT PAMPERIN

In June 1959, Robert Pamperin was diving for sea snails off La Jolla Cove in California when his companion, Gerald Lehrer, heard him scream for help. Turning, Lehrer saw his friend upright and unnaturally high in the water with his mask missing. As he swam closer, Lehrer watched Pamperin slowly disappearing into the crimson waves and, diving beneath the surface, he saw his friend being dragged to the seabed in the jaws of a 22-foot shark. Scouring the water for his remains, the US Coast Guard found only a single swim fin.

5) PACIFIC COAST ATTACKS

Four shark attacks occurred over a 15-day-period off the Pacific Coast of the United States in 1984 – beginning with the gruesome death of a 28-year-old abalone diver. Omar Conger was resting vertically in the water and looking out to sea when his companion, Chris Rehm, saw a great white shark rear up out of the water behind him. 'It grabbed him from behind, and while shaking him violently, pulled him under the water,'

Rehm later told researchers. The shark then resurfaced and released Conger, swimming straight at his companion. Rehm pulled his friend onto their dive mat and swam ashore, but Conger bled to death in the water.

6) BARRY WILSON

Swimming near Lover's Point off the Californian coast in 1952, 17-year-old Barry Wilson was seen by onlookers to jerk suddenly from side to side. The young tuba player then screamed before witnesses saw a shark rearing out of the water to attack him from the front and drag him under. Wilson resurfaced seconds later, screaming and flailing his arms in a pool of blood. Five fellow swimmers fought for 30 minutes to drag him back to the beach through the rough surf, but he bled to death before they reached the shore.

7) TERRENCE MANUEL

A ten-foot white pointer ripped off Terrence Manuel's right leg in 1974 as he struggled to scramble into a boat driven by his friend, John Talbot. The 26-year-old had been diving for sea snails in 30 feet of water when he suddenly burst through the surface and shouted 'Shark!' Talbot rushed to save his friend but was unable to prevent the attack and was instead forced to watch as Manuel bled to death in the water.

8) RANDALL FRY

Cliff Zimmerman was diving for abalone with his friend Randall Fry off the coast of California in 2004 when disaster struck. Zimmerman reported that he turned from Fry for a millisecond before hearing a 'whooshing sound' and feeling the water move 'as if a boat went by'. He spun around to see Fry gone and a shark fin surfacing momentarily before the surrounding water turned red. Zimmerman then swam for his life and his companion's severed head and body were found separately the following day.

9) RODNEY FOX

Rodney Fox was defending his Australian spearfishing title in 1953 when a great white shark grabbed him round the middle and dragged him through the water upside down. The predator released him as he gouged its eyes, but soon returned and attacked again. Fox jammed his arm down the beast's throat and pulled it free again, ripping the flesh from his arm. The shark released him and then returned a third time, dragging Fox along the ocean floor. After nearly drowning, the teenager was released and was pulled aboard a nearby boat with his ribcage, lungs and upper stomach exposed. Miraculously, his main arteries remained intact and he survived after four hours of surgery and 360 stitches.

10) BROOK WATSON

The first recorded victim of a shark attack, British merchant sailor Brook Watson was swimming in the harbour of Havana in 1749 when a shark attacked him once and then came back for more. The 14-year-old's crewmates saw the attack and dragged him from the bloody water, saving his life. Despite losing a foot to the shark and later having his leg amputated, Watson went on to serve for nine years as a Member of Parliament before becoming the Lord Mayor of London.

The *Daily Telegraph* article is obviously the work of somebody who knows almost nothing about sharks: it is clearly a 'cut and paste' piece, if only because the last victim mentioned, Brook Watson, lost

Photo: Gérard Soury, courtesy of Walter Bernardis

a foot, which could hardly be rated as severe. Thousands of others have lost their lives to these predators since records were first kept.

Indeed, in one holiday period in Margate, South Africa, in 1957 — that ended up being labelled 'Black Christmas' — there were several fatal shark attacks over a period of weeks. Recent history has many examples of what these and other 'man-eaters' can do to the unsuspecting. This unprecedented series of shark attacks spread fear the length of the coast and led to thousands of holidaymakers abandoning in panic the coastal resort towns and villages in scenes foreshadowing those in the movie *Jaws* decades later.

Cumulatively, with six attacks in a comparatively short time, this rogue shark activity led the authorities to respond by calling in the South African Navy frigate SAS *Vrystaat* to depth-charge the sea around Margate. Hundreds of dead fish floated to the surface after the underwater explosions — but no dead sharks.

Interestingly, author Al J. Venter was a crew member on board that former Royal Navy Whitby-class warship when all this happened: a lowly rating, he clearly recalls the crew regarding these depth-charging events with considerable scepticism.

His skipper, Captain Terry-Lloyd apparently shared those sentiments, but then orders are orders …

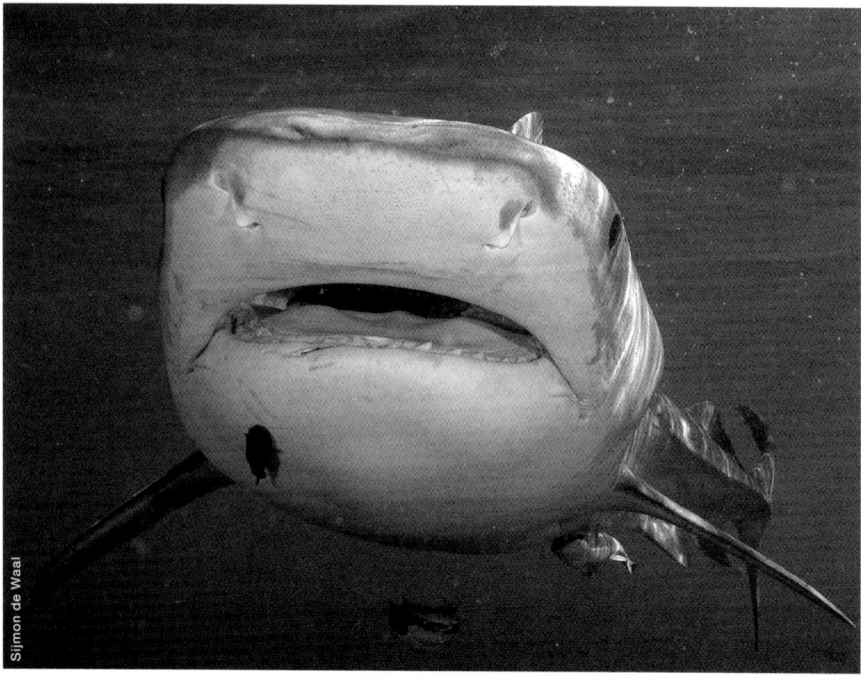

Sijmon de Waal

More recently, there have been several more shark attacks off Port St Johns in what was once called Transkei. The last involved Zama Ndamase (16), a provincial surfer for the Border provincial team who was in the water with his brother and other members of the local surf club at Second Beach in January 2011. It was the fifth shark attack in that area in three years, four of which were fatal. Young Ndamase bled to death before help could be summoned.

Then again, going back a little further, we are also aware of six shark attacks – three fatal – that were reported around the Cape Peninsula between 2003 and 2005. Two of these occurred in deep waters and four in False Bay.

An interesting insight into these events emerges from the International Shark Attack File (ISAF) which, in 2006, investigated 96 alleged incidents of shark-human interaction occurring worldwide. Though this is mainly historical data, the attacks do provide something of an insight into how and why these attacks happen and give some background to the nature and possible cause of attacks.

'Upon review,' declares ISAF, '62 of these incidents represented confirmed cases of unprovoked shark attack on humans.

'"Unprovoked attacks" were defined as "incidents where an attack on a live human by a shark occurred in its natural habitat without human provocation of the shark".'

It goes on: 'Incidents involving sharks and divers in public aquaria or research holding-pens, shark-inflicted scavenge damage to already dead humans (most often drowning victims), attacks on boats, and provoked incidents occurring in or out of the water are not considered unprovoked attacks. "Provoked attacks" it says, usually occur when a human initiates physical contact with a shark, for example a diver is bitten after grabbing a shark, a fisherman while removing a shark from a net, as well as attacks on spearfishers and those who might be feeding sharks.

'The 34 incidents not accorded unprovoked status in 2006 included 16 provoked attacks, five cases of sharks biting marine vessels, two incidents dismissed as non-attacks, two "scavenging" incidents, and nine cases in which insufficient information was available to determine if a shark attack was involved.'

While these details might be regarded as mundane, they certainly put matters into perspective with regard to sharks not being the wholesale killers or attackers that many people believe they are. In fact, experience has shown first, that most sharks are wary of people in the water and second, that sharks are generally *not* the aggressive monsters the media makes them out to be.

In fact, from personal experience, this writer has found the opposite to be true, with a few exceptions, of course: after all, we are dealing with predators.

An interesting article by Brian Walsh appeared in an edition of America's *Time Magazine*.

Headlined 'Farewell to Sharks (And Yes, That's a Bad Thing)', Walsh maintained that Steven Spielberg had a lot to answer for, specifically with regard to sharks.

He goes on to say that 'his greatest sin may have been the damage he did to the public image of sharks. His 1975 megahit *Jaws* didn't just usher in the era of the summer Hollywood blockbuster; it indelibly imprinted the concept of the shark as a killer, as the enemy of man (And John Williams, who wrote that chilling theme music: you're not blameless either!).

'People who had never so much as waded in the ocean became convinced that sharks were a menace, better off dead. As a kid paddling in the New Jersey surf – where I was probably more likely to encounter medical waste than any shark – I know that's how I felt,' Walsh stated.

Walsh also suggests that, in reality (unlike in the movies) unprovoked shark attacks are extremely rare, and fatal ones even more so. According to the International Shark Attack File, only six people worldwide were killed by sharks last year.

'But human beings haven't returned the favour. Each year, fishermen kill as many as 73 million sharks, usually cutting off their fins – which are valued for shark-fin soup, a popular dish in Asia

– before tossing the bloody carcasses overboard. Tens of millions of other sharks are likely to die each year by accident because of fishing gear set for other species.

'As a result, the International Union for the Conservation of Nature estimates that as many as a third of all shark species are threatened or near threatened with extinction, including the great white. Sharks aren't the true killers – we are.

'Part of that terror' says Walsh, 'may actually stem from a famous attribute of sharks: their need to keep swimming in order to breathe. While that's not actually true for every species of shark, many do need to employ what's called ram ventilation in order to respire, swimming forward with their mouths open, letting the water – with its oxygen – flow through the gill slits.'

As Eilperin writes, 'This is one of the reasons people see sharks as scary: cruising along as they display their sharp teeth, they look as if they're poised to attack at any moment.' What appears to be a prelude to aggression is just a poor shark trying to catch its breath.

Shark attacks occur under all sorts of conditions. A woman was attacked a few years ago in the United States by a bull shark (zambezi shark in southern Africa) while standing in 60 centimetres of water. She looked down and saw a shark of less than two metres swimming away. Yet, she still needed 40 stitches, even though nobody has been able to say for certain what kind of shark it was.

Gary Haselau, a Cape Town diver who, years ago, spent time with Jacques-Yves Cousteau on his oceanographic research ship *Calypso*, recalled a November day in 1942 when he was walking along Clifton's popular Fourth Beach, watching two swimmers making for the shore. He saw 18-year-old medical student Willem Bergh in the waves close to the beach where hundreds of people were relaxing and sunbathing. Within 20 metres of the shore this young student was attacked by a shark that Haselau estimates at more than seven metres in length, undoubtedly a great white.

'Both his legs were severed with one bite,' said Haselau, 'and we all watched horrified as the shark slowly cruised out to sea with the head and arms of the dead student visible above the waves in its jaws.' In years to follow Haselau himself had some serious brushes with a variety of sharks, but counted himself lucky he was never faced with a shark like the one he saw that day at Clifton.

Gary was hired as a photographer by the Cousteau team and spent quite a bit of time on board the *Calypso*. It was while on a visit to the

Caribbean and diving with the crew that he had a fairly serious run-in with a tiger shark that attacked the divers for no apparent reason whatsoever.

Shark attacks in South Africa have dwindled in recent years and in the past two decades there have been only about a dozen reported fatalities although, whenever one does occur it hits the headlines, like the one which resulted in the death of Ian Hill off Pringle Bay in the Cape, at the end of 1997.

In reality, there could have been quite a few more people who have disappeared while swimming or diving and whose deaths have never been fully explained. I know of at least two divers – both spearos – who we presumed were attacked by sharks but were unable to prove it. Scuba divers, in contrast, are much more fortunate: the number of shark attacks on that fraternity worldwide in the past half-century, I can count on one hand.

Today, sharks have a more sympathetic profile among divers and the public generally and in 1991 South Africa became the first country in the world to pass laws making the great white shark a protected species.

The mood might have changed but the shout of 'Shark!' is still one that can give anyone who hears it a bad case of the shivers.

Sijmon de Waal

CHAPTER SEVEN

'SUICIDE BY SHARK' AND OTHER STORIES

Jerry Buirski looks back on a lifetime of diving with sharks. Much of this activity took place in the days when it was still possible for spearfishermen (and women) to kill sharks with Bangsticks, before they themselves ended up on the predator's menu. Indeed, those were some of the most exiciting times. Struis Bay, a two-hour drive east of Cape Town, was where it always happened and divers who visit the place today continue to encounter sharks in the water. Jerry tells it like it was.

THE GREEDY 'RAGGIE'

For eight years I was a full-time commercial fisherman, sometime professional diver, spearo, shark hunter and, occasionally, wreck searcher and part-time maritime salvager. Much of this activity took place in the waters off Struis Bay – or more commonly, Struisbaai – in the South Cape.

In the summer months we'd go out early mornings and haul in yellowtail on handline or spinners. We'd return for lunch and head out again later in the day and spearfish. Bad weather and cold water made the sport unpleasant during winter months, so income from May to September would be erratic.

During those months I'd work on other things like diving for sinkers and, from old wetsuits and bicycle inner tubes, I'd fashion 'finger-protectors', for those who fished with handlines. I would sometimes trade galjoen for meat with some of the farmers in the area and would occasionally dive in their farm dams to look for lost pipes or pumps. Among my professional diving buddies, I did my image irreparable harm by once borrowing a knitting machine from an old lady and produced 60 sweaters.

Those were tough times. All I could afford was the cheapest equipment available. My shooting rigs were nothing like the fancy stainless steel and carbon equipment used by today's generation of spearos, the majority of whom only dive over weekends and holidays.

Because we were in the water almost every day and in all sorts of conditions, we often managed to recover lost gear. One of the happiest moments was when I became the owner of a brand-new stainless steel spear that was lodged in five or six redbait pods on Five Mile Bank. After breaking away the pods, it served me well for many months, both for shooting gamefish, as well as bottom species and also in provincial trials and national competitions.

Which all goes to prove that divers are the ultimate improvisors, because most times they simply have to be. Which, I suppose, is why we survive some of the trickiest and most implausible situations imaginable while at sea.

On one occasion, diving on the Bank, I saw a rather large raggedtooth shark watching us from one of the rocky projections on a reef. It was obviously taking a lot of interest in our activities. From experience, we'd learnt that sharks would arrive in the water around us as soon as we'd shot a few fish. Then they'd come in close and more often than not try to steal our catch. This was the same tactic, incidentally, used by great white sharks. We were not unduly concerned this time, but obviously this raggie was up to something and it ended up being worrying to us because this was a particularly large critter of about three metres.

Having worked one area, we returned to our boat, headed up-current and started another drift.

It wasn't long before I'd gone down and managed to shoot a very nice yellowtail (*Seriola lalandi*) on the bottom. I immediately began to head for the surface in order to play the fish from there with my floatline. The water was clear but patchy and a bit dark in places because of cloud. I was also aware that it was impossible to see the ocean floor from the surface.

As I started to drift away in the slight current, it felt as if my fish had become snagged on the bottom, which was also not unusual because a speared yellowtail will often seek shelter under a rock or in a hole in the reef or a wreck. Undettered, I again swam down my floatline, all the while keeping it taut. Which was when I spotted a large shark twisting about on the ocean floor with just enough white from the yellowtails's stomach to tell me that I was on target.

Only then did I see that the head of the yellowtail had been bitten off and was drifting to one side. Obviously, this meant problems because the shark had devoured my fish, spear and all.

As I again began to surface, I found that my buddy had joined me from the boat to check on developments. I hadn't quite reached the surface when, like a bullet, he shot right past me and hauled himself straight into the boat.

Once I came alongside, I handed my floatline to Dave Foyn and told him to pull the shark up: it had gobbled both fish and my spear, so it was 'attached' at the other end of the line. Dave also gave me his speargun with a .44 Magnum banger already fitted. His well-crafted protection devices are still the envy of many South African spearfishers.

With the raggie four or five metres from the surface, I saw that only a short section of my beautiful spear still protruded from its jaws. Dave allowed the shark to swim in a wide circle at that depth and each time it passed me, it kind of chomped in my direction, almost like an angry dog restrained by a chain.

I finally managed to dispatch the shark with a clean shot to the head and we hauled it on board, but not without a lot of effort. The only way to retrieve my spear was to open the shark completely, which meant that I also got my fish back, or what was left of it.

We took some pictures and one of them is still my favourite diving photo ...

MY FIRST GREAT WHITE

The first time I encountered a great white shark in its own mileu was at Agulhas Pinnacles. It came during a dive with André Hartman when we were spearfishing a mile or so offshore and opposite the lighthouse. The Pinnacles are distinctive for several reasons: they reach up from the bottom to within about five metres of the surface at two places, there is always fish around and, as I was to discover, there might be sharks as well.

We were operating from a flat-bottomed boat, only four metres long and painted with a camouflage pattern that had been loaned to us. Normally it was used on inland lakes to hunt ducks and geese. So strictly speaking, it was not seaworthy. But we were broke and hungry, so what the hell ...

Obviously unsuited to the kind of open waters off Struis Bay, the craft had no electronic equipment and not even a compass. Still, André and I were old hands at the game and had worked this area for many years. We were familiar with enough landmarks in the vicinity to enable us to dive on five steel wrecks. But it had been a while since we'd last visited the Pinnacles and weren't entirely certain which prominent stretch of bush should or should not have been in line with the lighthouse. Also, with continuous development, coastal features tend to change and bush is often cleared.

Once we reckoned we were close enough to drop anchor, I started to kit up, the idea being that while André enjoyed his routine pre-dive cigarette, I'd search the shallower part of the reef. It is interesting that André Hartman is the only world-class spearo I know who smoked, which was unusual: he was not only a first-class diver, but apart from achieving international status, had been South African spearfishing champion several times.

Once in the water, I reckoned I would configure the lay of the reef since certain fish are found in specific areas, a lot of it dictated by depth. Once overboard, I swam in a wide circle in a bid to find the shallowest spot. I managed several dives but this was no shallow dive, even if things looked promising because there was a lot of undersized fish around.

Finally, after I'd surfaced again and looked towards the boat, all I could see was André doing a Watusi rain dance on the narrow deck, his long thin legs in his black long johns flapping in the breeze. He was obviously trying to attract my attention and since I am quite deaf, I hadn't heard him calling me when I'd surfaced earlier.

While gyrating on the boat, he pointed excitedly, extending his arms and making chomping motions with his hands. He was actually trying to tell me that there was a large shark in the water. Having caught my eye, he gestured that I should head back to the boat *pronto* ...

This is the kind of thing that happens often enough in Cape waters even today and spearos who venture out must be prepared for just about anything. In those days, most of us would have our Bangsticks, as they are known in the trade, stuck under our left wetsuit sleeve.

Effectively this is a short metal tube into which has been inserted a cartridge, usually in .44 Magnum or .357 Magnum calibres. The Bangstick is screwed onto the tip of the spear that is normally used to shoot fish. With the 'banger' in place, the speargun would then be fired in the normal manner. Ideally, it would be aimed at the threatening predator's head. On impact, the sharp end of the spear would strike the cartridge primer and all being well, that would be that.

I'd been alerted by André that there was a large shark in the waters around the boat – and though I hadn't seen it yet or had any idea of its size — I removed my 'banger' from under my sleeve, attached it to the end of my spear and swam slowly towards the boat. I was careful not to splash my fins and took the additional precaution of screening the waters all around me. If the shark was going to hit me, I wanted to see it before that happened so I could at least try to protect myself, which was what the 'banger' was for.

It's worth mentioning that one of my diving buddies had commented often enough in the past that when you're in the water with sharks, your eyes became like those of a chameleon: you tend to acquire an observation arc of 360 degrees ...

I was still out in open water, with André reversing the boat towards me, hauling the anchor rope tight, when I suddenly realised that the shark had swum up right alongside me. It was a big mother, metres long. For a while it lay motionless in the water with me firmly fixed in its vision. Seemingly undeterred, though actually very worried, I just kept swimming to a position where André could haul me over the low gunwale.

He told me afterwards that he first saw the shark when it swam underneath the boat, but then I was already some distance away. It didn't help that it took him a while to catch my attention. Throughout, I was being circled by a monstrous great white shark and I wasn't even aware of its presence.

We hung around a while, but decided that since there was little current at that location, we'd move along a stretch to 'safer' waters.

André got the engine going and I started hauling in the anchor rope. Halfway through, the anchor snagged something on the ocean floor and wouldn't budge. We all but stood that little boat up on its bow but neither of us could get it loose. Literally, we were stuck fast.

After debating a possible course of action and not having seen the shark again, I told André that I'd kit up once more, go overboard and free the anchor. It was our only option and my buddy didn't argue. Rather me than him, he joked ...

Once in the water alongside the boat I pulled hard on the anchor rope until it was vertical in the water below me. With that I took a deep breath and headed down, hand-over-hand down the rope until I reached the bottom. The anchor chain was caught around some rocks and though I tried to unravel it and keep a watch for the shark, I wasn't successful. The anchor was about as snagged as it had ever been.

I headed for the surface again and told André that another dive would probably do it and I started to breath-up again. Just then this huge shark was suddenly right alongside me. Again! It appeared to have materialised from nowhere and had probably watched the entire first phase of me trying to free the anchor chain without my being aware of it.

One giant kick of my flippers and I was back on board the boat again.

'It's your turn now, André.' I told him. He gave me the kind of look usually reserved for the village idiot, coupled to a wry laugh and promptly lit another cigarette. Only after he had finished his smoke and the shark had disappeared again, did he slip over the side.

I was still looking over the gunwale at his floatline disappearing into the murk when I heard something rattling on the deck behind me; it was the anchor and André was already back on board! He'd freed the anchor in record time. With dirty water everywhere, we eventually anchored close inshore and ended up recovering more than 50 kilograms of sinkers, as well as a lost weightbelt.

The lasting impression of my first great white was that it was one of the most beautiful creatures imaginable. Sleek, powerful and graceful in its movements, it was something remarkable – and quite terrifying – to behold, especially from up close. No question, I was overwhelmed by its presence.

I even recall its great black eye observing my every movement: and its slightly open jaws, although André was of the opinion that that was probably not a good sign.

LET'S GO FOR A RIDE

In the bay in front of Struis Bay there are several different reefs. Most are quite flat and one has to swim about a good deal to find what we divers like to call a 'hotspot', which is where it might be possible to shoot a sizeable fish or two.

After you'd bagged a few, it is customary to continue the search, the boat accompanying you and staying close enough for the 'bak-kie-boy' to keep an eye on what's happening in the water around him. The divers, in turn, string their fish onto a float that is attached to a stainless steel stringer, but it is only done after we'd established that the booty was in fact dead, and for good reason. The first rule among spearos is that a struggling fish attracts sharks.

In these waters at the southern tip of Africa, from about January onwards, there are often shoals of small hammerheads in the area which more often than not bite chunks out of our booty. We'd usually counter that by boating our catches instead of stringing them. Still, I'd sometimes get back to my float and find the entire catch and stringer missing, most times without having felt or seen a thing.

This culprit was almost certainly a larger critter, probably a great white. It always astonishes me how quietly these large denizens can move about the ocean, sometimes getting within an arm's length of a diver without betraying its presence. For this reason we never use steel traces, because once taken, the shark might head off into deep water, not only with our fish, but also our entire floatline. It is not impossible that it would drag the speargun along as well, something we might have been customising for years.

There came a day when I encountered a great white shark that didn't simply creep around when there were divers in the water.

It happened off Saxon Reef, on the Struis Bay side of a large exposed stretch of granite called Bulldog. Quite often there would be seals milling about in the sheltered water to the leeward side of it; this was before I discovered that to get to where the ellusive musselcrackers found sanctuary, I had to swim underneath a stretch of breaking waves behind Bulldog.

We had already been at sea for quite a while and the other guys – Charlie Shapiro, Mike Keulemans and Dagh Calitz were a bit bushed and taking time off on board. Erik Lombard, also a former international-class spearfisherman, was keen for a dive, so we kitted up in our wetsuits.

Once in the water, we finned about and started to work the reef, making sure to keep clear of each other so we didn't spook the fish we were trying to hunt. After about half an hour or so my stomach started to rumble, so I returned to the boat and used the opportunity to dump my fish in the hatch.

Back into the water again, I found that Erik was some distance away from the boat and that he had quite a string of fish drifting from his float: just the motivation I needed. I swam over to where I had last been and started hunting again. Within minutes I had a stroke of luck: somewhere in a deep hole I shot the biggest roman ever! It was a significant trophy, especially since it proved difficult to get it out of its sanctuary.

Having strung the roman, I kept on diving. At one spot I was drifting on the surface when I felt a strong tug on my floatline, followed almost immediately by another. Both surprised and peeved – I thought that the roman might possibly still be alive – I was still debating my next move when my speargun was suddenly almost ripped from my hands. It was a powerful effort and my body was jerked to one side. I surfaced immediately.

As I looked out of the water towards my float, all I could see was the entire rear end of a huge great white shark thrashing out of the water from side to side. There was no question that my prize roman was in its jaws.

I yelled loudly towards the boat and the reaction of all on board was immediate. Charlie moved directly to the controls, Dagh made for the anchor while Keulemans jumped up and down. Meantime, the shark was towing me towards deeper water when it suddenly stopped and everything went quiet. Waiting for the boat to reach me, I used the opportunity to fix a banger onto my spear.

Suddenly the shark took off again, this time like a creature possessed. With me in tow, it shot through the water at such a speed that I had difficulty holding my mask against my face, all the while clutching the speargun with my other hand.

Though my mask was half-full of water, I was close enough to the shark to see its tail thrashing about at the end of my floatline. That went on for perhaps a minute, when things quietened again and the boat picked me up.

We went over to where Erik was still in the water, totally unaware of the drama, and picked him up as well. He actually wanted to go on with the dive as he'd been doing well with the shoot, but once we told him about the shark he was aboard in a flash.

So we all sat around on the drifting boat and pondered our next move. A short while later, with my floatline still in the water, I decided to haul it in. On doing so I could see that my buoy had been almost bitten through by the shark.

THE COMPETITION SHARK

Struis Bay has always been a popular venue for Boland spearos to hold their annual provincial trials. For quite a few years it also made sense for Western Province to compete there because it was easily accessible and fish were plentiful. But then again, so were sharks of all species.

One Saturday morning was such an occasion. There were five or six boats in the water, all trying their luck on Five Mile Bank. For the duration of the competition I was with Piet Vuilwater on his boat.

Piet had the unfortunate habit of enjoying his time in the water completely alone. It is something that is never encouraged, but he would head out through the breakers or from the harbour and who but he was to know that he would be diving that day? And when that happened, he usually would go into the water with his speargun tied to his boat, which would then also be his float. It is difficult to argue against such logic, though obviously there was a real element of risk.

This time though, we dived as a pair. We had only been in the water a short time when we had to move our location in order to find more bottomfish that would count towards our competition points.

Somewhere along the way, I shot an immature banksteenbras. As I headed towards the surface, I hauled the fish closer in order to kill it with the diving knife that I always wear on my left forearm. Moments later, an impressively large great white shark came in from nowhere and started to circle me, obviously attracted by the still struggling fish.

I immediately let go of my gun as well as the fish and started to swim directly up the floatline, the idea being that the shark would remain below with its next potential meal. But that didn't happen because the shark headed straight up the floatline towards me. Meantime, I was empty-handed and still some distance from the surface.

So I started hauling on my floatline and when the speargun came past the shark's head, it gave a sudden jerk and swam off. This allowed me enough time to reach the boat and signal Piet that there was a shark in the water with him.

As he was hunting directly below his boat, I had to wait for him to surface and then join me.

Now the signal used by divers in the water to indicate that there is a shark in the vicinity is to hold a fist up in the air. Moreover, if that diver has his head underwater, it also means that he is looking directly at the shark. That usually necessitates the boat approaching cautiously to make the pickup, and the boat therefore has to come up carefully and as close as possible, so as not to hit the diver.

I was still considering my options, when there was a sudden mad scramble on board the boat nearest to us. As if by some hidden signal, all of its divers got out of the water as quickly as possible. It was obviously the same shark and were its actions not quite so aggressive, it would have been quite comical, because it moved from one boat to the next, chasing all the unsuspecting divers in the area. It was a noisy spectacle, one group of spearos laughing at the expense of others, but after a while we pulled our boats close together to decide the next move and decided to finish the trials closer inshore. That would be around the wreck of the *Pioneer*.

GLAD TO MEET YOU ...

During the course of one recent winter, I was to help Martin van der Merwe restructure one of his earlier boats from the bare hull up.

The weather was outstanding and the ocean was flat calm, with pods of whales cavorting in front of Traill Witthuhn's house, which was where we worked because of his expertise and tools.

Finally came the day when the boat was finished. With a fresh coat of bright red paint it looked splendid and sea trials beckoned. The problem just then was that the weather had turned and the sea had become dirty. Visibility was down to a few metres, which is hardly ideal for spearing fish. We waited a few days and when things didn't improve, we decided to take the boat out anyway. We'd head out to the *Pioneer* wreck and see how the craft performed in medium seas.

Like most spearos, I had several wetsuits, the majority in various stages of disrepair. This time I chose a very old camouflage suit that had originally been given to me by Zululand diver Mark Anderson some years before, after we'd hosted him in our local waters. It wasn't an ideal setup because the wetsuit was coming apart at the seams and some of the white backing had come loose and was flapping in the water as I moved about. Elsewhere, the neoprene was literally coming apart. Still, I thought the suit was worth one last dive ...

Having reached the wreck, we first dived around the *Pioneer's* protruding engine-block, but there was little going on in the water.

We went across to what we refer to as the kop of the wreck, which lay to the northeast, and there we did about two dives apiece. Having come to the surface, there was little to do but gripe about dirty water.

I told Martin that the excursion was worth a final dive. If there was still nothing to shoot, then we'd head back to shore.

According to Martin, he was 'breathing-up' when, out of the corner of his eye he thought he spotted the white flaps on my wetsuit approaching him in the water. As he turned towards me to see why I had come so close, he was confronted instead by a great white shark that was easily five metres or more long. The water was so dirty that he could see only about a metre of the predator at any one time, but judging by the size of its head and its girth, this was a monster. When the beast's tail eventually moved past, he surfaced, called the boat and quickly climbed on board.

Meantime I was still in the water, unaware of what had taken place perhaps 15 metres from where I was hunting. I'd dived down deep and somewhere in the grey-blue soup I managed to shoot a wildeperd, otherwise known as a zebrafish. As I drifted slowly towards the surface, I hauled in my catch and dispatched it with my knife. At about that moment, I reached the surface to find everybody on board the boat, Martin included, and they were all looking at me.

When that sort of thing happens in these waters, you don't have to be told that there is a shark. I knew it even before they gave the customary 'chopping hands' signal. I did a quick 360 degree scan in the water and finally got into the boat, never having seen anything.

The encounter clearly illustrated the reality that there are a lot more sharks in the water – great whites especially – than most divers *and swimmers* realise. It is also true that they are only regularly spotted when the water is clear. Were that not so, there would almost certainly have been many more attacks than those that have taken place.

While the perception among the public at large is that sharks are killers, my experience had proved that within reason, they are actually incredibly curious. Most times they will make an approach simply to see who or what you are, and observe what you are doing. That done, they invariably continue on their way.

This kind of encounter is commonplace. I have lost count of the number of times that people in the water with me run into great whites, while these sightings elude me, even though we are often only metres apart.

FIVE SHARK SPECIES IN A MORNING'S DIVE

Every November in the South Cape, just before the wind starts to blow and school holidays are imminent, the spearo community banks on good weather and clear blue seas in deep waters, somewhere past the Five Mile Bank.

Since there are no shallow spots for yellowtail to congregate, we use working birds, like gannets and terns, to point the way to where the fish are shoaling.

On bright sunny days, the fish usually feed mornings and late afternoons, invariably with birds looking for pickings. During the middle of the day we would make other plans, or possibly dive 'blind', which can be tiresome and no fun at all.

At all times Martin and I would take along our rods and try some of Traill Witthuhn's tin spinners in the sea. We would sometimes surprise ourselves by catching a fish or two. So we started trawling these spinners or, in the lingo, *rapalas*, while heading straight out to sea, often pulling them along behind the boat. A strike would be the signal for the divers to go overboard and tackle the shoal from behind. The boat tender, meantime, hauls in any catch made on the rods.

This has always been a very successful ploy, to the extent that there were many days when we would be the only boat to return with a full quota of fish. The downside among these bluewater shoals was all types of large sharks accompanying them – in effect, the shoals are their food supply.

One red-letter November morning we encountered five different species of shark during the course of these dives. These included a great white, a mako or two, some zambezi sharks which usually prefer warmer waters, bronze whalers as well as dusky sharks.

Of them all, the two with the 'worst' reputations – great whites and makos – were the only maritime predators that did not always attempt to steal our catch. The biggest culprits were bull sharks, more generally known in these waters as zambezi sharks (and notorious as so-called 'man-eaters'). When there were struggling fish in the water, the zambies would head in like terriers and grab all that came before them.

This was fine for the first 20 metres of clear water, when you could see what was going on. Below that depth, the water was a dark and often murky haze, the kind of territory which zambezi sharks prefer. It is no secret that further up the coast, there are invariably zambezi sharks at the mouths of South Africa's rivers where the visibility is often zero. It is there, too, that some of the worst shark attacks take place.

Off Struis Bay, yellowtail would be hooked on our lures, we would go overboard as the shoal approached and I would usually allow the others to shoot first so that I could check the surrounding water. That would give me the opportunity to get close to the shoal and choose the biggest critter for my target.

About then, things would become interesting. There would be three or four struggling yellowtail in the water when a half-dozen zambezi sharks would suddenly appear, moving at speed. They would swim straight up to the struggling fish and each of them would try to grab one. These sharks could be extremely aggressive. They would sometimes come between two divers on the surface and steal their fish, literally from out of their hands.

One of our regular divers was Danie Raubenheimer, who is no friend of sharks. In fact, he has a loathing for them which might also be interpreted as an innate fear of the creatures. The first time he dived with us and a shark posse arrived, he threw his speargun away and tried to climb on my back. To avoid shark problems we decided to move to a point further out. Meantime, we again put out our lures.

The result was that the process was repeated: fish would grab our lures, the divers would go overboard, shoot more fish and before the haul could be taken on board, the sharks would arrive.

It was a regular jumble of floatlines and fish in a tangle, clambering on board, jokes about being taken by sharks and more than an occasional curse when the bastards got our fish. Eventually though, we'd end up with a good haul of fish and enough anecdotes with which to regale our wives or girlfriends.

We soon realised that it was pointless trying to kill all the zambezi sharks in the area, because there were just too many of them. When they really started to arrive in an area where we may have been working, it was much easier just to leave that area completely and hunt elsewhere along the coast.

There were times, of course, when we'd work a shoal of yellowtail and manage to shoot a number of fish without being bothered by zambezis. When this shoal eventually swam off, we'd roll our floatlines onto our spearguns and drift apart so that when the boat returned to pick us up again, we'd board from both sides.

On one such dive, Martin and Danie were already on board, having shot their quota. As I swam up, Dave Foyn was on my right, also heading towards the boat, but with his head out of the water.

It has always been a custom of mine, prior to hauling myself over the gunwale, to give the surrounding water a final underwater scan. This time, on turning around in the water, I found a very large great white shark lying motionless right underneath my dive buddy Dave! The shark's great head was probably less than a couple of metres below his torso, but he swam on, oblivious of the threat, his head clear of the water.

I promptly lifted my own head out of the water and screamed 'Blue!'. Dave turned, observed me pointing at him, put his head under the water and saw the shark. If he had been casually finning before, he raced through the water those last few metres and, as the boat approached, the shark slowly turned away and swam off.

Fairly often we'd encounter mako sharks in these waters, easily the most beautiful shark in the ocean. Such confrontations would usually only take place in warm, blue waters, well away from the coast.

The mako is blessed with the most beautiful large eyes and a bright, almost 'metallic' blue back. We'd often find it swimming just below the surface, sometimes moving from one diver to the other and leasurely checking them out before moving on again. According to the guide books, it is something rare to make contact with a mako while diving, but we spotted them just about every summer when conditions were favourable.

MY CLOSEST SHAVE YET ...

Several of my diving friends, and in particular some of the older and more experienced ones who had achieved Springbok colours, have survived shark attacks. Some were bitten quite badly in the process.

Yet, in spite of the many encounters with sharks in my eight years as a commercial spearfisherman – never mind the years since as a sport diver – there was only one encounter that could have ended badly for me. Another great white some years later might have vacillated between attacking me or not. But that time I took no chances and it ended up on the ocean floor feeding a multitude of crabs and crays.

My most dangerous experience occurred while I dived at Puntjie, about eight kilometres east of the Breede River mouth. It was a weekend excursion of the kind we made often enough each year, only this time we overnighted on somebody's farm. We were given a rustic hut in the middle of a dung-strewn field and were woken the next morning by some cows around the hut licking the salt from our wetsuits.

The other divers in the group were Martin van der Merwe and Dagh Calitz, the only member of the party who knew how to exit the extremely narrow river mouth through an even narrower channel blasted through the rocks. Since we were unfamiliar with the reefs, it took us a while before we found a decent area with enough large fish to shoot.

I have since come to know the area quite well: all the Cape Town aquariums' poensies were caught there.

On this Sunday, our last day of the outing, we managed to find some large underwater reefs that were covered in marine growth that included huge seafans and pink hard coral. The surface water was quite dirty, but it was clean underneath, though a bit gloomy because the sunlight wasn't able to penetrate the gloom.

Personally, I've always enjoyed diving in dark, clear water, since most fish would then come much closer than otherwise. And since I'd been struggling for several days to quickly remove my anti-shark 'banger' from my sleeve, I'd fitted a short length of braided nylon to the device and it was hanging in a loop from under my sleeve. After several turns in the boat as a dive tender, it was Dagh's turn to keep watch. I went into the water, well aware that Martin was somewhere on the other side of the reef, swam down several times and came up empty-handed.

Then, having kicked off from the bottom after again finding nothing worth shooting, I was suddenly confronted by a great white that was easily three metres-plus, moving fast and heading over the reef directly at me, all the while shaking its head vigorously from one side to the other, its classic mode of attack. Obviously this was an unusual event and I watched the predator with great interest. The moment it came to a point directly below me, it turned and kind of 'stood' on its tail, it's sharp snout pointing upwards.

The diving community, and particularly us spearos, has had enough experience with great whites to know that these sharks usually attack their prey directly from below. They do it with such force that they often lift their entire bodies – with their quarries in their great jaws – right out of the water, exactly as they do when they breach with seals successfully targeted.

All these images ran through my mind at that moment, well aware of the next step. Which was why I quickly turned around in the water and jerked my speargun and body down towards the shark. As expected, the tactic had the required effect and the great beast turned away and swam in a circle below me, still moving very erratically. Remember, this was a breath-hold dive and I was still some distance below the surface: at some stage soon I'd need a fresh gulp of air ...

But I did have enough time to slip the banger out from under my sleeve and to load it onto my spear. If the critter came for me a second time, I knew I would be ready for it. I was also aware that I would have to finish this confrontation as soon as possible. To get to the surface again, I'd have to go through a stretch of cloudy water with almost zero visibility, and if it came for me in that dirty stretch, it could be tickets.

Still circling, the great white came round towards me again. This time I went straight for its head and once more it turned away. By now my options were becoming critical: I needed air! So I just pointed the speargun between its eyes and pulled the trigger.

The shark was so close to me that when the banger exploded between its eyes, the spear had emerged from my speargun by only a few handwidths. But the shark was truly hit and for a brief moment I watched as it corkscrewed towards to the bottom.

It was difficult to replay the entire episode in my head afterwards and accept that the entire encounter probably took less than a minute, from start to finish. During this time, a thousand thoughts flashed through my mind. As for holding my breath, that was never a factor: my body must have found some air somewhere, probably in my legs ...

Martin was the first to hear the banger go off and he swam down to have a look at the shark. It lay on the bottom, upside down, it jaws fastened onto a chunk of reef. A little while later we shot another spear into its tail and hauled it onto the boat where we tried to tow it to Puntjie with our floatlines. I wanted a photo, but my camera was lying in Dagh's pickup at the slipway where we'd launched and our destination was miles away.

Taking it ashore proved a huge effort. We were towing the creature by its tail and its pectoral fins acted like two huge water brakes, apart from it being exactly the same length as the boat. Our tow rope snapped three times. Twice we managed to get it attached again, but while going through the surf zone into the river mouth — which was coming down with a strong flow of dirty brown water — the ropes broke for the last time.

Many years later, after I had started to work at the Cape Town Aquarium, I met Tinus Beukes who had been holidaying at Puntjie for many years. He recalled a great white washed up on the beach with a spear in its tail. There is no question that, had I not spotted the shark approaching, I would probably not have survived an almost certain attack.

We've been diving in that part of the coast long enough to also accept that great white sharks in the bay are more aggressive than elsewhere. There have even been several ski-boats attacked and holed over the years.

During a national spearfishing championship in the area, one of the competitors lost a flipper to a great white, an incident that took place close inshore. Also, former Springbok Attie Louw survived his second attack on the other side of the bay.

A contributing factor might be commercial fishing trawlers that shelter in the bay during westerly gales. These boats are often followed by large sharks that feed on the offal and fish thrown overboard.

ENCOUNTERS OF THE "BLUE" KIND

Quite a few years ago, while still living near Agulhas, just south of Struis Bay, I helped with the organisation of the South African National Spearfishing Championships being held there.

Since it wasn't possible for me to take part in the trials, I made myself available as skipper for one of the teams invited from abroad. This should have been the New Zealand team, but when they pulled out at the last moment, I was shifted across to look after the Portuguese contingent.

It was soon apparent that sharks featured prominently in the minds of just about all the overseas competitors. Obviously, if the competition was to be a success, we would have to do something reassuring and attempt to shift their mindset to the many species of fish likely to be encountered underwater: not an easy task, with shark stories galore floating about.

Some of the competitors had another good reason for concern. During the first day's diving on Saxon Reef, in truly atrocious visibility, a member of the Zimbabwe dive team looked around and saw a shark right behind him with its jaws open. Almost instinctively, one supposes, the spearo kicked out at it. The raggedtooth shark – for that is what it was – retaliated by grabbing one of his flippers and pulling it right off his foot.

We managed to retrieve the flipper afterwards and judging by the almost circular holes made by the raggie's teeth, we were able to assure those concerned that it wasn't actually a 'man-eater'. When I later showed members of the Zimbabwe team photos of several types of shark, they immediately pointed out the raggie. Which didn't prevent the *Cape Argus* the next morning from running a sensationalist article about a spearo almost being savaged by a shark at the South African National Spearfishing Championships.

It didn't take long for us to become good friends with the Australian and Portuguese teams. Once the main events were over, we even took them to Gansbaai to dive for crayfish and perlemoen.

Wolfgang Leander

A member of the Aussie group was Rob Torelli, editor of an Australian spearo magazine. I gave him some information on sharks, which he later passed on to Terry Maas, who eventually used some of it in his excellent book *Bluewater Hunting and Freediving*. I am quoted as follows:

'Most white shark encounters do not result in an attack,' explains South African bluewater hunter and commercial fisherman Jerry Buirski. Struis Bay, South Africa, well-known for white sharks, is a favourite dive spot for Jerry, who over the years has had enough encounters with whites to categorise them into types.

'The first is when you meet the shark, usually at the surface, but when the boat arrives the shark has already gone,' he says. 'After getting into the boat for a few minutes, you go up-current, resume diving and you never see it again.

'Second is when the shark stays with you, swimming up and down on one side while you head for the boat, which has to approach very close.

'Third might be the most disconcerting. This is when you are on the bottom and the shark appears, usually from behind and always from the left side. It follows your ascent, often pointing its sharp snout straight at you, seemingly without swimming at all. This angle to the shark's head makes it impossible to get a kill shot with a banger (Powerhead). If the water is dirty, the tip of the shark's jaws will be literally inches from your speargun, which you will hopefully have turned in its direction. The encounter usually ends with the arrival of the safety boat.'

Jerry explains that the white shark population off Struis Bay fluctuates annually and for no apparent reason. He feels comfortable enough these days with white sharks in the water, so much so that he has actually shot and recovered yellowtail right next to a white shark – and he has done so more than once (these South Africans are tough!).

From his own experiences with whites, it seems that once you have seen the shark and it knows you have seen it, you are safe, Jerry says. There is a *caveat* here: Jerry goes on with the story of five Australian spearfishermen who were diving together. Four of them saw an approaching white and it did not attack them. But then it went for, and attacked the fifth diver, who never saw the shark.

Wolfgang Leander

There are several references in the book that mention Tommy Botha, who is South Africa's most famous spearo. He had the following to say about great whites: 'One thing that amazes me about these great creatures is their ability to sneak up on you in the water without being seen. At Struis Bay we normally dive close together when we hunt yellowtail. There will be four guys looking in every direction but when you see one, it's normally already on top of you.'

'SUICIDE BY SHARK'

Long before I came to accept that great white sharks were not nearly as bad as they were supposed to be and that their reputation was based largely on hearsay – though, to be fair, there have been people taken by them – I had an experience that confirmed that belief.

It was towards the middle of a summer (December/January) and peak spearfishing time in these southern waters, that I experienced one of those personal catasrophes that we all go through at one stage or another: I got myself unceremoniously dumped by a woman, two weeks before the wedding.

This was quite a shock, after almost getting hitched after so many years of living alone, and over the next week or two I went shark crazy in the water. The depression I was under was so powerful that I decided to see if I could get a shark to put an end to it all.

Obviously, my dive buddies were aghast. Quite a few of them regarded me as having gone 'over the edge', which I suppose I had, in part. Whenever we encountered sharks, even a great white, I'd stay in the water longer than the rest of the guys in a bid to actively confront the predator. After that happened three or four times, my companions were convinced I'd gone loco.

On one dive, on a perfect day, the water was about as filthy as it gets in these parts. But just past the lighthouse and fairly close inshore, we discovered a patch of beautiful clear water. Fortunately, the wreck of the triple-expansion steamer, the *Camphill*, was right there, so we put a buoy down and kitted up. Since clear water on this wreck is a rarity, and it seems to attract some really big fish, I was especially keen to go down, and as quickly as possible.

However, there was a strong current running, and on the very first drift, a great white came swimming past quite leisurely. It barely spared us a glance but I nevertheless immediately fitted a 'banger' and went after it. But then it turned away and disappeared into the distance.

Because of the shark, we'd entirely missed the wreck, so we got back on board the boat and went up-current to restart the drift dive.

Now, not only was there a shark in the water around our boat, but experience had long ago taught us that when that happens, it is dangerous to jump in, simple as that! But intent on my personal quest to get eaten, I simply ignored the threat and promptly went into the water again.

Once the mass of bubbles around me dispersed, my first impression was of a large great white shark swimming right alongside me. It was a little below my own depth, but close enough to brush against my flippers. Hoo-boy!

With the protection of my 'banger', I immediately gave chase, but that didn't have the required effect either. The shark took off at speed, though that didn't prevent it from turning its head and peering back at this crazy human trying to come after it.

Meantime, André Breytenbach had joined me in the water, sensing somehow that perhaps the shark was not as dangerous as he'd earlier believed, and that I possibly needed 'looking after'. Martin and Dave, however, wouldn't budge. They stayed on the boat and watched our

Morné
Hardenberg

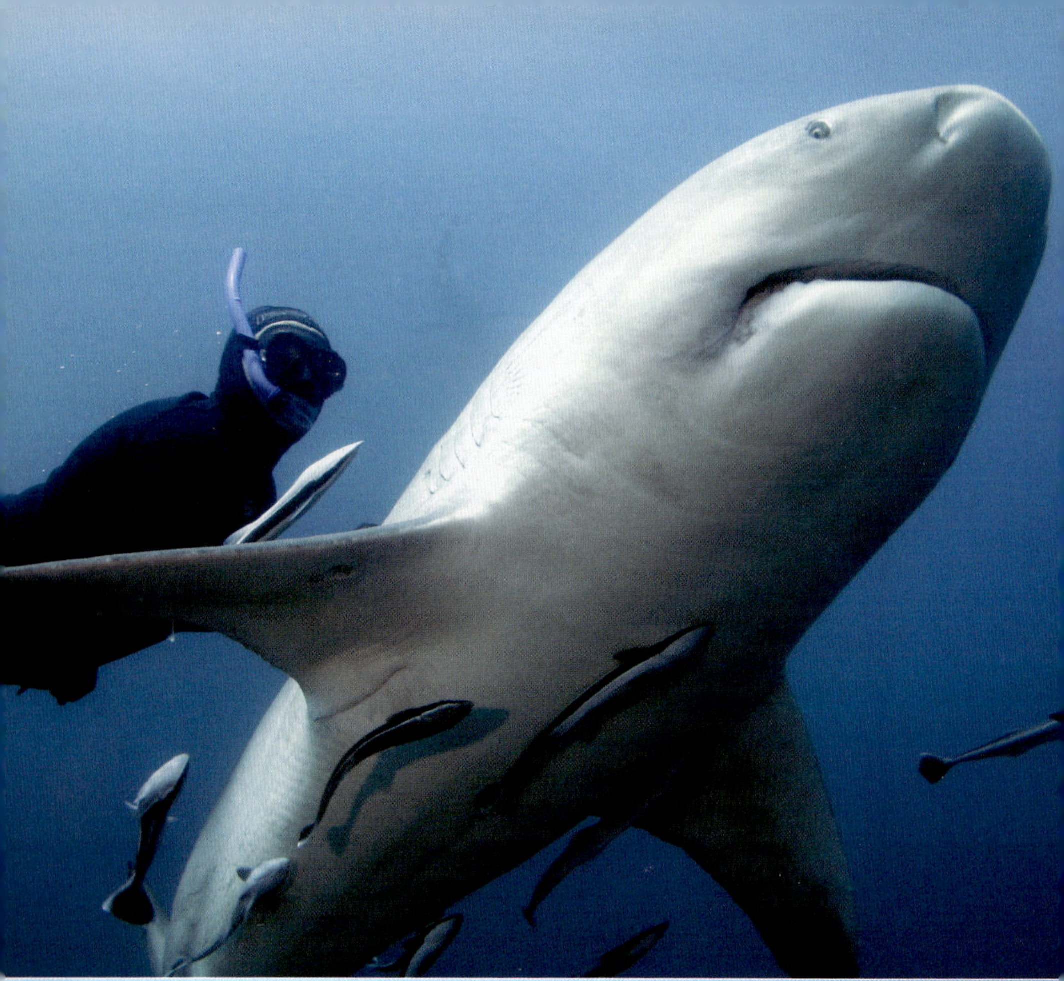

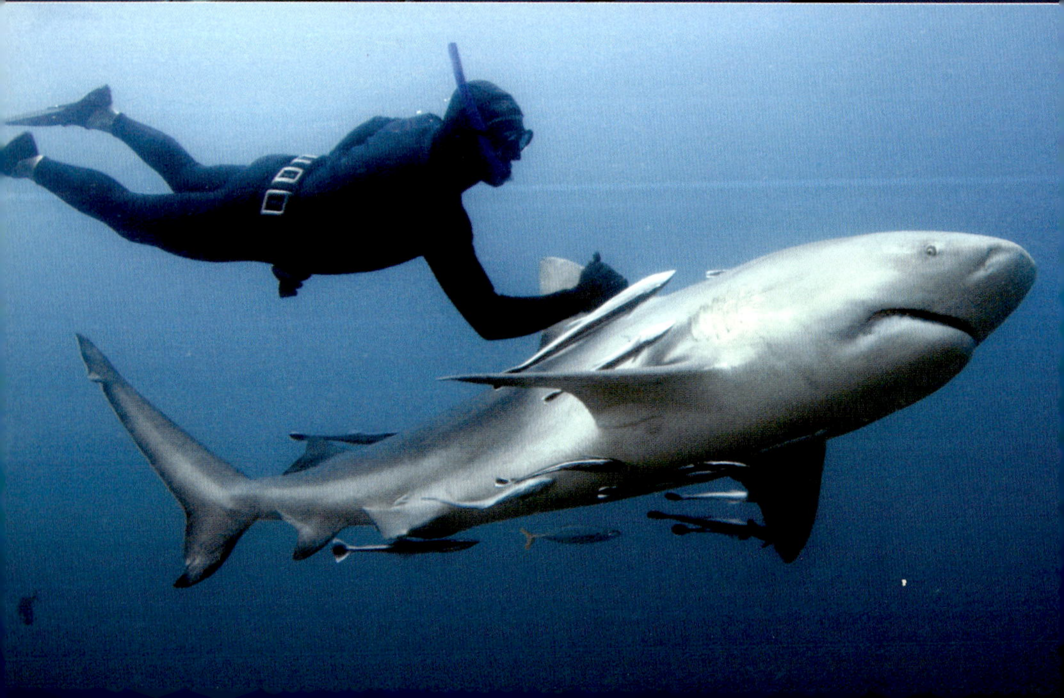

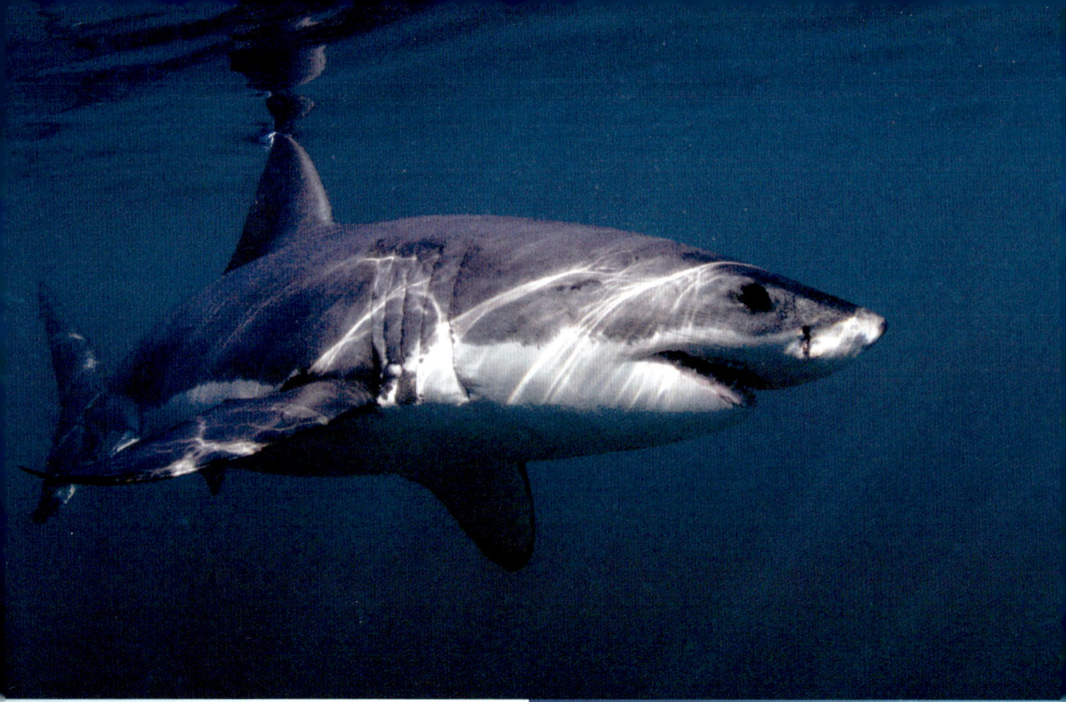

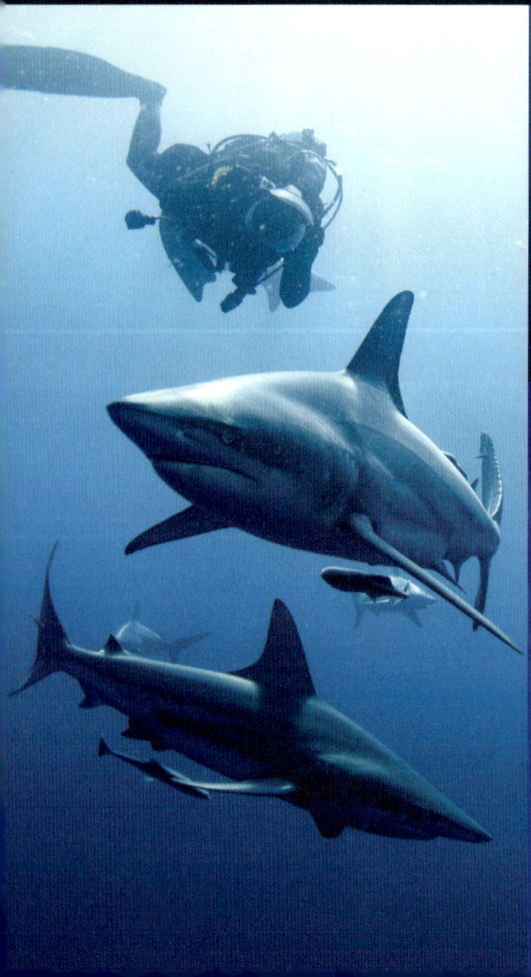

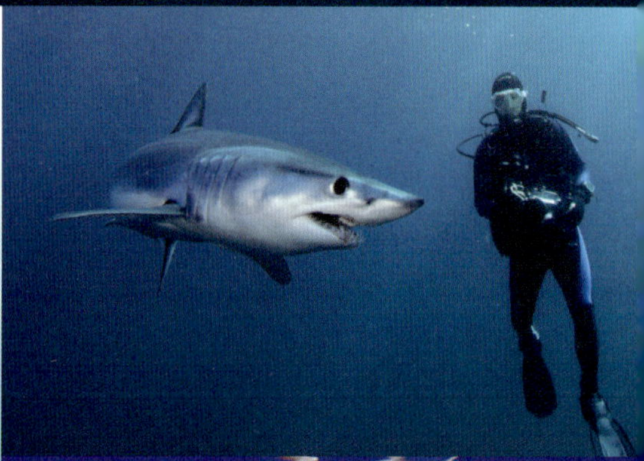

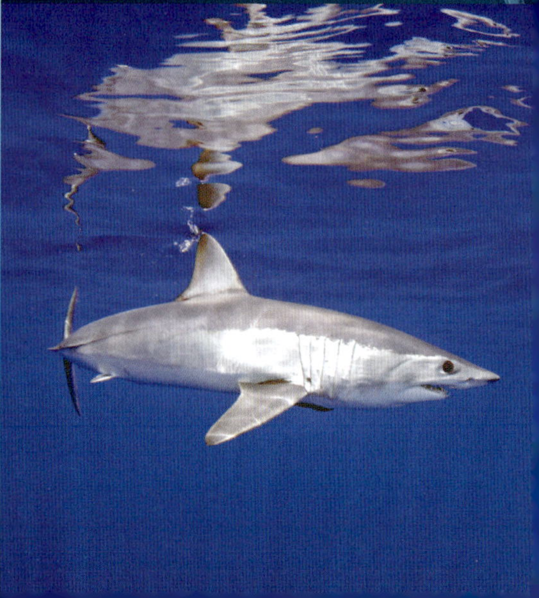

every move, fully expecting the shark to reappear at any moment, they told us afterwards.

With the shark having spoilt the only patch of clear water, the others said they were eager to leave. However, I insisted on one more drift dive. As I told them, 'I want to teach that bugger a lesson'.

Again I jumped into the water and once more, after the bubbles had cleared, there was the shark, right between my flippers. This time it was even closer than before.

Still eager for the kill, I went after it, but unlike before, the predator didn't disappear, even though I ended up chasing it all over the ocean. Finally André grabbed me from behind and insisted that I get back onto the boat.

On the way to the harbour, after shedding my kit, I replayed my moments of madness. Turning the events over in my mind, I realised that twice, when I had literally jumped almost onto the shark's back, it could easily have made a meal of me had it wished to do so. The fact that it didn't might have been construed as a warning, which, just then, I thought I should heed.

And so I did.

SHARK TALES UNLIMITED

For most of my commercial spearfishing career, I would regard people like Tommy Botha, Erik Lombard, and especially André Hartman as the ultimate shark divers.

All three were hugely experienced open-water spearos, with Tommy at his best rated among the top two or three international competition divers. André went on to help make a number of shark films for *Discovery Channel*, the BBC and others. Of them all, I always believed that Tommy probably had the most great white experiences. He was twice attacked by 'tommies', as he liked to call them.

Imagine my surprise when Terry Maas' book appeared and I read that Tommy had had about 30 encounters with whites. By then I was already way past the 50 mark, having stopped counting long before that.

Since all of these professionals had been attacked by great whites at one time or another – and survived – I took a great interest in their comments. Erik once warned me that while I was apparently laid back with regard to possibly being targeted, my attitude would almost certainly change after I'd had my first serious run-in with one of these critters. Fortunately, this has never been the case. While most of my diving buddies would go straight home after being threatened by a white shark, more often than not I would usually keep on diving. I'd do so by simply moving my boat to another reef.

Which was how, on a single morning's dive not long ago, we encountered five large great whites on five different reefs in the South Cape.

On one occasion I'd overheard Martin telling a bunch of overseas divers that I had the distinction of having had the most free-diving great white run-ins in the world. While I had never thought of it like that, it could very well be true. South Africa has long been acclaimed as shark country, with Struis Bay home to possibly the most great whites in southern African waters.

It was only a few years later, long after shark cage diving ventures in the vicinity of Dyer Island had taken off to cope with thousands of tourists who were eager to experience these creatures in their own milieu, did André Hartman and Mike Rutzen leave me behind in the number of free-dive shark encounters experienced.

The greater hammerhead was probably the last of the larger sharks with a fearful reputation that we encountered in our area over the years, but that didn't happen too often. We'd sometimes find them when diving on the Agulhas Pinnacles, another attraction for some of the larger fish we were hunting, especially in summer. It was also a place where we'd been shadowed by great whites often enough.

There was usually a current running at the Pinnacles and customarily, we'd drop off on the seaward side of this beautiful reef and try to do as many dives as possible before being picked up. After such a drift, we'd usually roll up our floatlines while waiting for the boat.

On one such dive, Martin had three smallish yellowtail on his stringer with the boat only a few yards away, when a monster of a hammerhead, probably six metres long, approached him and tried to take away his fish. His stringer was right next to him, so there was no missing the shark's intent.

A minor struggle ensued and while Martin tried to hang onto his speargun, the shark's 'shovel nose' became entangled in the floatline and stringer. Moments later, an alarmed hammerhead shark took off. Martin had no option but to let go or risk injury.

By this time I'd become aware of the commotion. I quickly swam closer in a bid to assist, when the hammerhead shot right past me, a line around its shovel, the float on its head and the three fish suspended just in front of its wildly snapping jaws.

A few minutes after we'd decided that it was perhaps best to return to the boat, Martin's buoy popped up some way down-current.

When we did eventually haul it in, his speargun was missing, the line obviously bitten off by the shark, though he did get two of his fish back.

To the north of Struis Bay is a large bay where we used to spearfish for days at a time without ever seeing a great white. The Western Province spearfishing union even held some of their competitions there.

Known as Die Mond, the area has acquired a nasty reputation for its unusual proliferation of sharks. For that reason, we only dive there if visibility is brilliant all the way from the ocean floor to the surface. Strangely, when conditions are that good, we seldom encounter sharks there.

Just around the point to the north is Arniston. From there, further to the east, there are some excellent spearfishing reefs, especially opposite Skipskop. Unfortunately, this is traditionally great white country, and for some years, we've found them resident in the area throughout the summer.

Once, I'd been wreck-diving off a large boat in the vicinity of Skipskop when I decided to do a little spearfishing: I'd been there before and knew that the bottom offered excellent prospects for bagging something on the seaward side of the wreck. The area was close inshore and I managed to shoot over 50 kilos of fish. Obviously, the other guys on board were impressed and they all decided to have a go the following day.

Unfortunately, just as the boat arrived at my favourite hotspot – I was in the bow with the marker buoy in my hands at the time – a huge great white appeared alongside. It was one of the biggest sharks I'd seen, much longer than the ski-boat from which we were intending to dive. After following it for a while and taking photographs, we reluctantly shifted to another location.

Traill Witthuln used to fish those waters and one time I went out with him and we caught a number of fish from his boat in about 15 metres of water. The catch was outstanding and included yellowtail, kabeljou, poensies and red steenbras. As Traill had something else on the next day, Dagh Calitz organised for Erik Lombard to take us all out in his ski-boat.

We were all excited at the prospect of the dive. Erik was fired up by tales of the big fish we'd bagged the previous day, and that meant that once on-site, nobody wasted any time before getting into the water.

In a flurry of gear, bubbles and excitement, it took only moments for the whitewater to clear. Perhaps three or four seconds later we found that we were right on top of what was possibly the biggest shark in the area.

It might even have been the original great white we'd spotted from the surface before ...

Sijmon de Waal

CHAPTER EIGHT

DIVING WITH TIGER SHARKS: A MOST UNUSUAL EXPERIENCE

I've dived with Walter Bernardis of tiger shark fame a dozen times and he always seems to have a fresh approach to his pre-dive briefings. Like when he warns those who dive with him, that they should listen carefully to what he has to say, because there is always an element of risk when there are sharks in the water with you.

We went out with him again on his boat, a few kilometres off Aliwal Shoal, in April 2011. Once over the dive site, he made the unusual comment about tiger sharks, as he phrased it, 'suddenly materialising from nowhere'. They were unlike any of the other sharks in these southern Indian Ocean waters, he told us.

'One minute there are only a bunch of blacktip sharks – perhaps a couple of dozen of them milling about in the water with you – and the next there is this huge tiger shark that has suddenly arrived from nowhere.' Moreover, it was not something that happened occasionally – it was, as he termed it, 'the shark's standard performance and, frankly, it can be pretty sneaky ... which is why we have to keep a wary eye out for these otherwise beautiful creatures.'

It's their incredible camouflage, he explained. The sharks would arrive and then, quite suddenly, they'd disappear again. A short while later, they would reappear out of the murk, but quite often from a totally different direction, and that kind of behaviour tends to unsettle people, he added. 'But once they're around, it's that much easier to keep track of them.'

'So keep your eyes open', he told us, 'and if you see one of these beasts, point at it so the others know it's there as well.' Also, he warned against individuals becoming separated from the main group. 'You do that and you create an interest, because you're isolated and these big sharks like to examine isolated things ... then you oblige me to leave the main group and haul you back again ... we really don't want that sort of thing happening on my watch ...'

Interestingly, Walter filmed a German cameraman who dived with him a couple of years before and who became separated from the main group. The enthusiast was so preoccupied with all the sharks around him that he tended to drift down towards the bottom, which lay a good 15 metres below the main group. Aware of what was happening, Walter kept the cameraman in view and after he'd spotted one of the tiger sharks moving in – a big critter, of about four metres – he went down, filming the scene himself throughout.

What you can observe from Walter's footage is the shark drawing close to the diver — who is aware of its presence by now and is also watching it carefully – and then the tiger shark turning on its side, open-jawed, and moving directly towards the diver's leg, but quite lethargically, which is something tiger sharks are known for.

'Of course he didn't let the shark bite him ... sort of shoved it away, but the experience demonstrated the level of threat associated with these contacts. What was important, stressed Walter, was that the entire action was slow-moving. 'It was kind of lazy, in fact ... nothing overtly aggressive.' 'Now, had that diver not been aware of what was taking place, or perhaps not even been aware that the shark was there – it could have come up from behind — it might have been a very different story,' said Walter.

He has himself been 'bitten' once by a tiger shark after he committed the cardinal sin that he always preaches against, which was when he turned his back on a large specimen he and his group had been diving with. He admits that he was lucky not to have lost his leg.

'I suppose you can put it down to overconfidence. I've been diving with these sharks for many years and this time, towards the end of a dive session, I'd seen the last of my group head towards the surface from the bait ball and decided to follow them to the boat.

'Then I felt a tugging on my fin and I thought, hello! All the divers are supposed to be ahead of me, so what is one of them doing tugging at my fin. I turned my head to see a tiger, with half my leg already in its mouth. My reaction in that millionth of a second was pure instinct as I jerked out backwards and it wasn't a moment too soon: the shark's jaws clamped shut with a thud and though my foot and calf were safe, its teeth still shredded a large section of my lower leg.'

A rather dramatic moment of truth during a baited shark dive off Aliwal Shoal: a blacktip shark bites the exposed hand of a diver in its jaws: one of the reasons why it is wise to keep your hands folded and under your arms when in the water with feeding sharks. The other photo shows the victim with his hand bandaged. Clearly, the shark could have caused a lot more damage had it wished to: it was obviously an accidental encounter.

What to do now? That was Walter's first reaction. Suddenly a fairly large flesh wound was spurting blood from dozens of jagged cuts and there were about 30 other sharks in the water. It wasn't red blood either, but green, which is the colour of blood at depth.

'I really believed that with all that blood around, I would have been in serious trouble, but the sharks simply ignored me, which was the first time that I realised that these predators are much more interested in fish than in humans. Now ... had I been oozing fish blood, I think it would have been the end.'

Walter has been involved in a few other incidents that involved sharks and human blood over the years, including two more incidents where he was accidentally bitten again, but nothing untoward happened and he still has all his fingers and toes.

'We had one fellow – also a German diver – during the sardine run last year, bit of a crazy fellow actually. We'd been diving near a particularly active bait ball along the Wild Coast and, as usual, there were sharks constantly emerging from the frenzy. I called time and told my group to pull back a bit, but this German guy with his camera just kept on filming, right there on the edge of the bait ball. I went to him and indicated that he should move back, but he ignored me, so what could I do but leave him to it.

'The next thing I knew was that a fairly large shark – a bronze whaler, I think it was, almost three metres long – suddenly appeared opened jawed right before him and a moment later it had all but enveloped the man's head.

'It was an accident of course, because the shark had been gulping sardines and its forward momentum had kind of shoved him right onto the cameraman. I actually thought we'd lost him, because again, there was blood everywhere.'

What had happened was that the shark's scissor-shaped teeth had penetrated fairly deep into the diver's scalp and also severed his nose, which was left hanging on his face alongside his displaced mask by a small thread of skin. Had it been an actual attack, the shark could have lopped off his head with ease, but it wasn't.

'We got the poor fellow on board, rushed him to shore and, to cut a long story short, he ended up being rushed to the Kingsway Hospital in Amanzimtoti by car – and that was five hours of sheer bloody hell – followed by a half-day session in the theatre with one of the top plastic surgeons in the country who stitched his nose back on again. One of the advantages of modern-day communications was that an entire surgical team was waiting for him by the time they wheeled him into theatre.

'Frankly, the man was lucky. But it also tells us all that you need to listen to those of us with good experience in these matters: if we sense danger, we're not being melodramatic ... it's a kind of gut feeling ...'

Since then, the German diver has been in touch with Walter Bernardis and recent photos show that he displays almost no scars from an otherwise dramatic incident.

Apropos Walter's own experience of having his leg almost re-moved by a tiger shark, Harvey Miller, a 36-year-old American div-er was snorkelling off Oahu's Bellows Beach in Hawaii that same month when, as a Honolulu newspaper phrased it, 'a large predator' – also a tiger shark – severed his leg above the knee.

This father of four had been snorkelling about 150 metres offshore while looking for turtles. He noticed some fish near him that looked spooked and then saw the shark's large, flat snout and felt a blow as the predator spun him around. The injury to his leg was enormous, but Miller was still able to punch the shark twice below its dorsal fin and that seemed to prevent it from coming in again and possibly finishing him off.

What also emerged afterwards was that a stranger, on hearing Miller's screams, waded into the sea, right past the shark and carried him to safety, a gesture that probably saved his life. Miller concedes that if not for that still-unknown individual, he'd have been a goner.

Notable too is the fact that the attack was the first known shark incident along a sizeable coastal stretch from Makapuu all the way around to Kaneohe Bay since 1958.

Prior to our dive, Walter spent a good 20 minutes briefing us before we kitted up and went into the water.

Meantime, his 'bakkie boy' chummed furiously, throwing buckets of foul-smelling fish and offal into the water to attract the creature. You could clearly see the oily gunge float away with the current and interestingly, it only took a few minutes for the first of the remoras to arrive. And, as Walter will tell you, if you have remoras, you will soon have sharks.

Basically, the bigger the remora, the bigger its host: so if you have a remora of about 40 centimetres, it means there is a pretty hefty tiger shark lurking nearby.

Wolfgang 'Wolfie' Leander, a septuagenarian shark diver and, after Walter, arguably the most enthusiastic tiger shark conservationist in the world, is constantly in the news about tiger sharks. He has been a moving force against the *indiscriminate* baiting of the creatures: while he is not opposed to divers viewing tiger sharks from up close, *how* they are actually attracted is the issue.

Wolfie was also one of the first to hail a new South African bait ball that does not harm sharks when feeding.

As he explains, people involved in the business of viewing tiger sharks from up close is the reason for the need to limit any potential damage to the shark's jaws, skin, teeth and its eyes. Previously, operators used metal containers (like drums taken from washing machines) to hold their bait, but this could cause injury, especially if the shark wrapped its jaws around the steel container or the wire rope holding it.

Consequently, he states, the system should have no sharp points, has to be 'soft' and preferably pliable. Ideally it should be of a resilient type of plastic so it will not break teeth if 'attacked'. It also has to be strong enough to ensure longevity within a hostile saline environment.

As a result of a good deal of research, the industry came up with what they termed 'ZIBS', a zero-impact baiting system, described by Patrick Douglas, CEO of Shark Diver, and others as a 'remarkable industry innovation'. It was the first of its kind anywhere and represented a unique and industry-generated evolution towards sustainable baiting practices.

Walter and others now use this device, which resembles a huge, oversized golf ball. It's most salient feature is that it does not allow the shark to get its teeth into it. Standing almost a metre tall, the ball incorporates a trap door, which can be opened by the diver master to access the bait and feed the hungry predators.

Over decades, Wolfie Leander has emerged as the accepted international spokesperson for the protection of tiger sharks.

When several individuals in the Umkomaas and Scottborough areas killed a bunch of tiger sharks, Wolfie was at the van of an international chorus of condemnation. He was all the more vituperative because one of the people involved in the slaughter was a commercial fisherman and the other was an Austrian national, who had recently arrived in South Africa, lived in Widenham and owned a company that actually offered tiger shark dives as part of his business. You try and make sense of that kind of skewered logic ...

Wolfie admits to having free-dived for more than half a century, but only turned his attention to tiger sharks in recent years when he first encountered them in the waters around the Bahamas.

'To swim with tiger sharks spoiled me completely – all other sharks I have encountered in the ocean, even large ones, pale in comparison.

'No other shark has a stronger expression of character than the tiger shark: The square head, the blunt nose, the comparatively large mouth (often tightly shut) and those inquisitive, intelligent-looking black eyes ...'

He goes on: 'I'll even go so far as to say that I haven't dived with sharks more gentle than the tigers of the undersea world. I feel reborn when I'm with them in the water ... I suppose you can say I'm hooked ...'

Last year Wolfie and his son Felix had what they like to refer to as 'the kind of friendly encounters with tiger sharks that we both love'. Father and son were in the Bahamas again, off Tiger Beach, diving freely with them.

'Felix put together a nice video clip. At the end of it you can see that one of the sharks was more than casually interested in Felix's camera.

'We were both at the surface when the shark, a huge female, grabbed the camera ... we looked again and she had it in her cavernous mouth. Felix held on tight and managed to get it out of the shark's jaws without hurting himself. It was actually quite easy as its bite was not at all that aggressive: in fact, it was rather gentle.

On diving tiger sharks in South Africa, Walter Bernardis offered the following opinion: 'Operators must not lose sight of what diving is all about, and that it is not just a meal ticket. Shark diving has earned a lot of money for cash-strapped local dive charters and carried them through several difficult years.

'More to the point, shark diving is something that is special: it requires time and patience to achieve results. It should not be prostituted into strict launch times, turnaround policies and shuttle-type diving — that is what reef diving on Aliwal Shoal today is all about, not shark diving.

'This lack of respect for clients and the dive itself will only lead to negativity in the international circles that tiger shark diving is a rip-off and that you don't see tigers.

'My message to all divers out there is simple: Yes, this is still the most rewarding dive of any description that you are likely to do, so choose the operator with whom you dive carefully and, more important, thoroughly enjoy the experience.'

In his pre-dive briefing, Walter offers numerous pointers, the first being that everybody needs to follow basic rules. If they do that, they are fine. But remember, there is always the possibility of an opportunistic bite that might injure a diver. So keep your arms folded, tucked away under your armpits.

He has had his high-pressure hose taken by a predator, the actual gauge. 'It might have been caused by its face glinting in sunlight, which, to a shark, might have resembled a fish. Another diver had had his air hose severed in a bite and that created a sudden rush of high-pressure air and caused him to immediately abort his dive. 'Which is why I warn about hands and arms ... flashing them about in the water attracts sharks, if only because they might mistake them for something edible.

'Remember too,' he continues, 'that while diving with tiger sharks, there are sometimes 20 or even 30 blacktip sharks in the water with you. These predators are sometimes two metres plus, so they are big, they are swift and to the novice they can look ominously dangerous, even if they aren't.

'Also, blacktip sharks are not afraid of divers. They tend to regard humans with them in the water in exactly the same way as they view other fish in the sea. Simply put, they don't regard us healthy beings as part of their food chain. Sick, or struggling, like any predator, maybe, but then you are not going to get into the water from my boat if you are not in good physical condition, which is why I demand a reasonably advanced dive qualification.'

Tigers are a different kind of animal, Walter warns. 'They *are* intimidated by human presence and it takes a while for them to become accustomed to us being there. And, as he always points out, scuba divers attacked on Aliwal Shoal are absolutely minimal: there are fewer than a dozen attempted shark attacks annually.'

So, he says, when you get into the water to be with these streamlined hulks of muscled cartilage that move gracefully through the water, very often within touching distance, relish the experience! Remember too, that sharks can move at astonishing speeds and their agility is quite remarkable: 'a shark can pass you at speed and turn right around within the length of its own body. "The ultimate killing machine", Jacques-Yves Cousteau called them.'

It is interesting that while this stretch of the South African coastline is historically regarded as zambezi or bull shark country, they're generally not part of Walter's entourage.

'We get the occasional zambie, but it doesn't hang about. I think the reason is that the bull shark is a hunter and not a scavenger. It's hugely aggressive though, as any diver who has spent time in these waters will tell you and from what I know – though I stand corrected – they're not attracted to carrion in these waters.'

'It is not for nothing that the zambezi shark has arguably claimed more human victims in these waters over the years than any other species of predator.

'I've been diving south of Durban since 1982. Even so, I still have much to learn about the behaviour of sharks, especially the bigger ones that have a somewhat deserved reputation for killing and sometimes eating people.

'I've been able to get close to some of the most dangerous creatures on God's earth, stroke them, ride their fins short distances and sometimes find them coming back for more. *And that is unique!'*

CHAPTER NINE

HOME OF THE ZAMBEZI SHARK: THE PINNACLES IN SOUTHERN MOZAMBIQUE WATERS

Barry Skinstad was a 22-year-old greenhorn when he first arrived in Ponta Malongane in Mozambique. He was well-regarded by his peers as an experienced small-boat skipper and spearo. Trouble was, Barry thought he knew just about everything that the ocean could throw at him, including sharks ...

As Barry tells it, he had behind him three good years of skippering and launching experience out of Sodwana Bay and had just acquired his very first dive boat; a six-metre Gemini named *Top Gun*, with two 85-horsepower Yamaha engines. By the time he got to the former Portuguese colony, he was eager for new adventures.

As he recalls, he'd heard through the grapevine that the camp at Ponta Malongane in southern Mozambique was looking for both a skipper and a boat. 'So, after meeting with Dave Kennedy, the facility's dive operator, I decided that this was a good chance to branch out into something new. Almost overnight, with some really interesting stories of plentiful fish and more whale sharks than I would be able to count – and fortified by who knows how many ales – I actually thought I'd discovered paradise.

'One of the tales that will always stick in my mind was reports of a resident whale shark off the nearby point. The word put out was that if you had the patience, this friendly, six-metre monster would eventually come to the surface and put its head on the pontoon for all to see. I needed no prompting, and to the consternation of my mother, I hooked my boat onto my pickup and headed north into

Mozambique, which some members of my family and quite a few friends believed wasn't the best move, in large part because the country was grappling with a pretty serious domestic insurgency. Mozambique at the time was distinctly 'war torn', as the newspapers of the day liked to phrase it and security was obviously a problem ... there were ambushes along many of the country's roads and people were getting killed. But I felt that it didn't really matter: In any event, I had made my choice.

My mother was right, of course, because the war in Mozambique's north was intense, but we saw very little of it so far south. The problem was that when she decided to visit me, she insisted that I escort her to and from the frontier, and, once there, we discovered an unexploded RPG-7 rocket deep in the trunk of one of the coastal milkwood trees near the beach. Truth is, as testified by an explosive projectile stuck in a tree, aiming accurately – either with a rifle or a rocket-propelled grenade – has never been something for which Africa was renowned. And so began some of the best years of my life.

My sojourn at Ponta Malongane began shortly after Mozambique's 1994 New Year celebrations and, being mid-summer, both the heat and the mosquitoes were intense.

The diving along that stretch of coast, and the spearfishing – my immediate interest – was good, but I soon found that it was nothing exceptional. In fact, we were working only one of several reef systems off the point at that stage, so it was not a vast improvement when compared to what I'd been accustomed to on Five Mile or Seven Mile Reefs in Sodwana. It was a disappointment initially, but then, like everything else, things change.

That came during my second week of working in Mozambique. Our base had originally been a campsite that for many years had catered for a variety of South Africans who routinely headed north across the border, usually with their boats in tow. They still arrived in small groups, most times looking for a place to pitch their tents and *braai*, and among them were a couple of Zululand ski-boaters always on the lookout for something different. A casual bunch, they would return to base before sunset and usually haul out the beers before putting their fish on the coals.

Afterwards, we'd all gather round and talk about the day's events. Our new-found Zululand friends had been coming to Malongane for years and they had a bunch of anecdotes about the 'old days', including one that immediately caught my interest.

During the course of our second or third evening of 'fishing stories', one of their party started talking about a rather unusual spot where the fishing was better than any of them had ever experienced. The catches were always good, he explained, and some of the fish they'd hooked were enormous, especially late in the afternoon. Some were so big, he reckoned, that they weren't able to land them in the boat, or, more often than not, the sharks would get them. There were hundreds of these predators in that area, they confided.

I was obviously intrigued. It was the first time I'd heard of a reef where there were sharks in abundance, so I pressed them for the location of this 'magical' reef and the reply came that it lay about four kilometres off the next point north of where we were mostly active.

My problem just then, was that only a short while before arriving in the former Portuguese colony I'd speared a large 'cuda while diving near Sodwana. As I was killing it, I accidentally slipped my index finger into its mouth and lost the tip ... these things can happen in a blink of an eye ... as it did with me. So, with all the stitches and bandages in place, I was forced to stay out of the sea for the first couple of weeks, though I was still sufficiently interested to discover what lay below the waves and that meant covering it in waterproofing a few times so that I could get into the water and look about.

My injuries didn't still my imagination after I'd listened to the Zululanders a few times. The prospect of diving among huge fish that had never before been encountered by fishermen was entrancing. I was dead eager and couldn't wait for my wound to heal.

After the next day's launch, I turned towards Dave and said I was going to head out and look around for some new reefs. I pointed the prow of the boat in the general direction of the place that had formed the basis of several evenings' wild tales and set out with two rods and a gillie, Thomas, my trusted boatman. We trolled for ages, all the while looking for something that might have resembled what I'd already dubbed the 'Golden Acre'.

We had no luck to start with, and were about to head home when I noticed some small birds diving a bit further out to sea. I would always call them 'tunny birds', because their actions suggested the presence of fish shoals, and, more often than not, tunny. It took us only a little while to reach the spot, when I suddenly realised that the blue waters around us had suddenly turned an unusual shade of golden green. Momentarily stunned, and with the screeching of two of my Shimano reels and a couple of rods bent almost double, I was immediately brought back to reality.

For a short while, everything on board was in chaos. Ten seconds later one of the rods flipped back into its regular upright position. But I still had the other rod in my hands and it was obvious that I'd hooked something that was both big and strong enough to peel line off my reel at a steady rate. A minute or so into the fight and there was a series of hard knocks on the other end of the line and almost immediately my rod went slack. The release was so sudden that I found myself on my bum on the bottom of the boat. I was completely dumbstruck and clearly I had no idea what had taken place, except that I'd lost two big fish in a row.

Never one to give up easily, and since there wasn't much current or wind, I turned the boat around and headed back towards the original green-coloured stretch of ocean. At the same time, I quickly took some onshore bearings so that I could return when conditions were more favourable. That was followed by several depth soundings. The first reading told me that my boat was hovering over a small pinnacle that rose up straight from the ocean floor to within 30 metres of the surface.

Still intrigued by the amazing colour of the water, I felt that hand injury or not, I couldn't rest until I had found out what had caused the sea to change colour the way it did. And because I always kept a spare mask in the boat, I thought I'd take a look. A short while later I positioned the boat so that it would drift back over the area.

Having hopped overboard, I wasn't at all comfortable in the water. I had no fins, no snorkel and no speargun. Meanwhile, Thomas constantly called out our position and I was finally about to find out what this remarkable phenomenon was all about.

Straining my eyes and feeling unusually vulnerable – the sharks mentioned by our Zululand friends were very much on my mind – I noticed a large shimmering mass some distance below the boat. I battled to comprehend what I saw, but as the boat drifted closer, a phalanx of huge golden heads rose up from the depths and I was confronted by hundreds of giant kingfish – *Caranx ignoblis* – which, almost in unison, rose up towards the surface and circled my boat.

I'd been diving for quite a few years by then, but this was certainly the most amazing sight I had ever seen. The shoal was enormous – so big in fact that I couldn't actually see through it. Nor could I make out where it ended. I'd speared one ignoblis off Sodwana in the 20-kilogram range, but these huge fish in the water around me were decidedly bigger. Then, as quickly as they'd arrived, the entire shoal disappeared into the depths and that apparition was replaced by dozens of sharks that were drifting in the depths immediately below me.

And that, basically, was how I came to discover what later became known as the Pinnacles in the waters of southern Mozambique. Geologically speaking, the rock formation was indeed a pinnacle rising off the ocean floor. Moreover, I had several good landmarks and I was determined to head back there in the morning.

There were no launches planned for the following day, and by eight the next morning *Top Gun* was hovering over the Pinnacles. With Thomas again at the helm and my hand firmly plastered, bandaged and wrapped in a surgical glove – I'd bound the ends with insulation tape to keep the water out – I was ready to go.

It didn't take me long to establish that the shoal I was after tended to remain on the current side of the Pinnacles. That meant I would be doing a succession of quick 20-minute drift dives: I'd complete one cycle, return to the boat and we'd head up-current again.

On my second dive as I levelled off at about 20 metres, I promptly came face to face with about eight zambezi sharks. For a moment or two I was completely non-plussed, but taking no time at all to regain my composure, I bolted for the surface. To my horror the entire pack followed me up. I don't recall exactly what came first — a gulp of air or the shrill bellow of 'BOOAAAT'!

Essentially, that was my brief introduction to spearfishing on the Pinnacles and interestingly, to this day its gives me a lot of pleasure whenever I take someone new to the place to do a spot of spearing: the fright, the occasional terror, or even the jaw-dropping awe – call it what you will — it is always the same!

Diving among a pack of aggressive zambezi or bull sharks can be a frightening first-time experience for anybody who has never done it before.

I sat on the boat for a long time after I'd emerged from the water that morning. I'd obviously survived, but I was still not quite sure what to do; sharks in the water is a serious mater, swimming with them on your own is even more so. The reality that there were more sharks in that location than I had ever seen in my life in one place before, coupled to the reality of everything promised by that wonderful Golden Acre, made the decision extremely difficult. After carefully considering my options, I thought, what the hell, and so I decided to try my luck again.

Almost gingerly I entered the water a second time. With the threat looming, I found that breathing-up was almost impossible because almost immediately about a dozen or more sharks started drifting towards me. I was about to get back into the boat when I again caught sight of the shoal of giant kingfish.

Taking my courage to its limits, I did a ridiculous dive and levelled out at roughly ten metres, not brave enough to go any deeper. To my delight the kingies swam straight up at me, which was when I put a spear into the first large fish that came into range. As the spear hit home and the fish tried to fight its way loose, everything suddenly changed; it seemed that every shark in the area was targeting my trophy.

Subjected to the pack instinct of a bunch of killers, the wounded kingfish disappeared in a cloud of blood and once again I was mortified, especially since one particularly large predator, with the head of my fish, together with my spear still embedded in its skull, started to head towards the bottom. Within moments my floatline was being whipped through my hands at a rate of knots.

Not thinking clearly, I grabbed the cord and tried to haul back my spear, only to find myself yanked violently downwards. Stupidly, the cord was in my bandaged hand and though I wasn't immediately aware of it, slicing into the finger that I'd earlier believed was well protected. Fortunately, another member of the pack bit through the line that held the spear onto the gun and I made my way back to the surface, blood pouring out of my finger.

That, basically, was the first fish I managed to spear on the Pinnacles. And it was also how I came off second best in the endeavour. But, and this was important, *I had the landmarks* and I knew that there were fish in abundance there.

Clearly, the location was something quite unique and once my injuries had healed, the next step was to do a couple of scuba dives on the Pinnacles to orientate myself to the basic structure of the reef.

The actual pinnacle covers a rather modest area, about the size of the average squash court, which was one of the reasons why so few fisher folk had discovered it in the past. There seems to be a ridge that runs parallel to the coast along the whole southern section of Mozambique and this pinnacle would be one small clump of rocks on this ridge; it rises up from about 45 metres on all sides to form a miniature table top.

At tides in the neap range it sits at about 30 metres, with a sheer drop of five metres towards the north and another on the seaward side. The inside edge has a more gentle slope with an enormous gorgonian fan dominating everything around it.

About 50 metres out on the seaward side, there is another ledge that also runs parallel to the coast, where, in the summer months, the diver is likely to encounter a large congregation of hammerheads.

From late October the fish start coming in, first a succession of small shoals, and, as the water warms up, they get substantially bigger.

It took a while for me to discover that the area is not only graced by giant kingfish, but that sea pike (both the blacktip and jello variety) as well as several other species of kingfish proliferate there; these include blacktips, yellow spot, blue fin and the ignoblis. There are also big eyes and bludgers, as well as kaakap on the bottom, together with the odd barracuda drifting by in mid-water. From June until August, you will also encounter some big wahoo.

What is clear is that the pinnacle is seasonal. Fish start gathering there in early November and move off again in March. But why these game fish arrive in such large numbers, still remains a mystery. Were it not for these shoals, I reckon that the numbers of sharks would be more modest. Indeed, more often than not, I would encounter sharks swimming with the shoals, waiting to pick off an easy meal.

Most of my time on the pinnacle was towards late afternoons, usually after work. We always found that from around three in the afternoon the pinnacle started 'working'. Where the fish went during the hottest part of the day also puzzled us: they probably headed out into deeper water.

In those carefree days I was invariably on my own, diving from my own boat and crazy about spearing fish. In fact, looking back, I reckon I spent three or four hours in the water at least three afternoons a week at the site.

Obviously, with time, familiarity with the sharks tended to breed contempt and I gradually got used to the presence of these critters in the sea with me. There were times when I dived down deep enough to see the bottom and could clearly make out anything between 30 and 50 of them. Zambezi sharks were the dominant species, but there were also blacktips, silvertips, hammers and occasionally some nice big tiger sharks in the three- to four-metre range.

Over the years I guess I became accustomed to diving with these predators, learning to read their body language and knowing when to get out of a bad situation.

Obviously, the Pinnacles simply had to be a unique diving location and with time, the word got out. Once the scuba-diving community started to arrive in numbers, more spearos also began to head towards southern Mozambique. The adverse side to all this was that some gung-ho enthusiasts began killing the sharks with Powerheads.

The first to go were the silvertips, the boldest of the shark community on the Pinnacles: they would usually rush in at you and take a lot of your fish that had just been shot. Occasionally, I'd see individual sharks swimming about that had not been very effectively targeted by these people, usually with spears still protruding from their heads. It was totally against the norm and, while there, I worked hard to prevent this kind of mindless excess. But most times, the thrill of the chase took over, especially among less experienced divers.

It took a while, but my objections did eventually filter out and the spearing fraternity took note. This, together with some stern words to a group of Durban divers resulted in the sharks being left alone for a while.

Things went smoothly for some years. Each season we would look forward to the summer and the big fish that heralded it. Good friendships were forged on the Pinnacles for me and life was good, especially after a successful 'hunt'.

Back on land towards evening, the divers were always upbeat, laughing and recollecting their stories of the day, and how they had 'miraculously' survived an afternoon on the Pinnacles.

Then catastrophe arrived with the longliners, usually Chinese fishing boats that played out kilometres of baited line with hooks across the ocean floor. They would drop a series of lines within a 20-kilometre radius and wait five or six hours before doing a series of relays to retrieve their catches. The toll of sharks slaughtered in this mindless manner was awesome and their numbers dwindled at a frantic rate. In fact, one year we never saw a single zambezi shark anywhere near the Pinnacles.

We tried hard to halt the carnage. On one of my trips, I actually came across one of these longlines and started pulling it up. Working from our modest-sized ski-boat, Thomas and I managed to release a dozen large sand sharks and two small hammerheads. It was clear that this kind of illegal fishing is astonishingly effective at trapping these creatures and the damage, as a consequence, is unconscionable. But then the Chinese have also been instrumental in the killings of almost all of Africa's rhinos, together with a significant proportion of the continent's tusk-bearing elephants for their ivory.

What seriously beats me is that shark fin soup is still sold openly in most Chinese restaurants in the West. Conservation groups have made a lot of noise about whaling, but are stunningly silent about this far more prevalent and destructive practice of slaughtering millions of sharks annually for their fins.

That is all they take once the predator has been hooked – the rest of the carcass – usually the shark is still alive – is thrown back into the ocean. I firmly believe that there is a special place in Hell for such cretins: a pox on them and their families ...

Obviously these criminal acts against nature could not continue indefinitely and we started to do something about it. After contacting government authorities at Maputo, we were directed towards the Mozambican Navy. They were eager to help but, they told us, they had no boat.

So we invited some of their officers to our site down the coast and offered to take them out to observe some of these excesses for themselves. A short while later eight naval personnel arrived, not only for a visit, but to stay with us as our guests. They were armed with AK-47 Kalashnikov carbines, Browning machine guns and, I was delighted to see, an RPG-7 rocket launcher.

Two days later, almost as if it had been formally invited into Mozambique territorial waters, a foreign boat appeared on the horizon. It headed straight for the Pinnacles. I loaded the troops onto my boat and we charged after the intruder. It wasn't long before we caught up with the vessel and tried to call them on the radio, both on our VHF sets and on 29 MegaHertz. There was no response.

On the fishing boat there was no missing our intent as we sped closer: just about everybody on the ski-boat was in full battle gear, frantically waving their arms in an attempt to get the putative poachers to stop. Once they saw troops and guns, the Chinese skipper promptly turned around and headed back out to sea. But he hadn't reckoned on the reaction of a certain Mozambique naval captain.

Our captain-friend wasted no time in ordering the sailor with the rocket-propelled grenade to fire at the ship. The man tried twice without success: the RPG charges had been doused with seawater on the hasty trip out, which generated a lot of spray and this infuriated the captain even more, especially since we were now abreast of the bastards. That was when he told his men to load and lock their AKs.

Moments later he gave the order to fire and hundreds of rounds thudded into the ship as the pursuers emptied their magazines.

At the end of it, the consensus among all of us was that it was a job very well done. An immediate consequence was that from then on – until the day that I finally left Mozambique – we never saw any longliners again in that area. Further up the coast illegal fishing continued unabated, but those criminals were effectively out of range of Maputo's authority.

Meantime, it took a couple of years for our dorsal-finned friends to return to their old haunt at the Pinnacles and, by the end of 2007, the numbers were almost back to normal.

Sijmon de Waal

CHAPTER TEN

SHARKS OF THE INDIAN OCEAN

Jeff McKay, formerly of the KwaZulu-Natal Sharks Board, looks at the activities of some of these predators, found in all the tropical oceans of the world. They are not usually a problem in warmer waters, like those encountered off KwaZulu-Natal, Mauritius or Mozambique, though it helps to understand the behaviour of different sharks that might be encountered.

'What about sharks?' In many dive resorts in Indian Ocean waters, the divemaster will invariably reply: 'No problem.'

Scuba divers in warmer waters rarely see sharks, partly because of the sites chosen by the guides, but also because divers are usually only aware of the inhabitants of the reef in their immediate vicinity. They tend to focus their attention on the reef bed and, except when ascending, rarely look at the waters above them. One South African dive group that went on an expensive Red Sea excursion in 2009, specifically to view sharks, did not see a single one of these predators.

Had they spent a couple of weeks diving raggedtooth sharks, zambezi or tiger sharks along the upper South Coast with Debbie Smith's 'Diving with Sharks' or Island Rock with Daryl and Clive Smith's operation, 'Mokorran Dive Charters', at Rock Tail Bay, they would have returned home with the memories of a lifetime.

The usually calm blue Indian Ocean abounds with sea life. Many species of fish – including sharks – share these waters and each of these predators has its place in the scheme of things. Indeed, all are necessary to maintain the ecological balance of the sea. They are not a great danger to divers and a sighting should be regarded as an enormous treat, rather than a threat.

The three most common species of shark seen by divers in more tropical waters of the Indian Ocean, are the blacktip reef shark, the whitetip reef shark and the grey reef shark. Other species might also be encountered, such as the sandbar shark, sicklefin lemon shark, silvertip shark, tiger shark, whale shark and the hammerhead. Game fishermen are much more likely to encounter pelagic species such as the oceanic whitetip shark, great whites, makos or blue sharks, but usually much farther offshore.

Remember, there are very few recorded shark attacks on scuba divers. In all the oceans of the world, in the past 20 or 30 years, sharks have killed only a handful of divers.

BLACKTIP REEF SHARK

Carcharhinus melanopterus

The blacktip reef shark is the shark most commonly found in southern Indian Ocean waters. This shark is aptly named as it has prominent black blotches, accentuated by a pale band on the tip of the first dorsal fin and the lower lobe of the tailfin. The other fins have black tips, and a white band runs along the flank from the anal fin to below the first dorsal fin.

The snout is blunt and rounded and the body is generally copper or grey in colour. Averaging less than 160 centimetres, the blacktip reef shark is among the smaller deep water or island sharks.

They are often found in the shallows near the shore on reefs or in the intertidal zone and large numbers are found in the shallow lagoons, particularly where the water is clear. Juveniles are often found in the shallows and the adults may patrol deeper water in Mauritius near drop-offs such as Casiers, a site frequented by divers from the Meridien Hotel. They are also found at Passe St Jacques, and as many as 40 sharks have been seen on a single dive at Flic-en-Flac. They are also common in Mozambique and in Seychelles waters as well as off the islands of Rodrigues, the Comores Archipelago, Madagascar and Réunion.

Those encountered in Mozambican waters have been found to be both swift and strong swimmers. They may be encountered alone or in small groups feeding on invertebrates and small fish, including kingfish and groupers. Most are inquisitive when divers first enter the water, but they often flee after a few minutes.

In shallow waters, they may become aggressive, particularly when they arrive in numbers or when there is spearfishing activity. This species has been responsible for a number of nonfatal attacks on human beings, probably because they mistook unprotected hands and feet for their ordinary prey.

GREY REEF SHARK

Carcharnus amblyrhynchos

When this shark feels aggressive, it warns you! It swims erratically, with its head up and wagging from side to side, back arched and pectoral fins down. That is a warning that you had better clear out – NOW! This is the classic body language of most small- to medium-size sharks.

Although the grey reef shark feeds mainly on fish, squid, octopus and crustaceans are also part of its diet. Spearfishermen have been bitten when grey reef sharks were in pursuit of their catch and when stimulated to a feeding frenzy, they can become dangerously aggressive. Grey reef sharks are also extremely curious creatures, and it is unnerving when five or ten of them approach within a metre of a diver.

Grey reef sharks are found in the western Indian Ocean, and they are abundant round all the islands including Mauritius, Madagascar and the Seychelles Archipelago.

They are mainly coastal, inshore creatures and inhabit fringe reefs and deeper banks from the surface and intertidal zone to depths of 140 metres. Juveniles are more commonly found in shallow waters and on reef flats. Adults, in contrast, generally patrol passes, rugged drop-offs and patch reefs in shallow lagoons where there are strong tidal currents.

They appear to prefer the leeward side of the island and congregate by day at particular places, often near areas where there are strong currents. Usually the sharks stay near the bottom, but occasionally they rise to mid-water or the surface.

Divers may encounter these sharks cruising in mid-water or see them leave the bottom and swim upwards to investigate their presence. They will approach divers in numbers, but soon disperse once their curiosity has been satisfied. After that they keep their distance and seldom reappear near the divers.

Although most commonly known as the grey reef shark (it has a grey body, fading to white on the underside), this shark is also known as the blacktail reef shark. The broad black band along the rear edge of the tail fin is an unmistakeable field mark. The second dorsal fin, the anal fin, the pelvic fin and the tips of the pectoral fins are black, but the first dorsal fin is never fringed with black.

It has a long, rounded snout, serrated upper teeth and a stocky body, and it is reported to reach a length of 2.3 metres, but most are about 120 centimetres in length.

SANDBAR SHARK
Carcharhinus plumbeus
This shark can be found in the tropical to subtropical waters of the East Coast of Africa and Indian Ocean islands. Divers most commonly find it in the 20–40 metre depth range.

The sandbar shark has little to distinguish it and is easily identified as a dusky shark. It has a short, rounded snout, a stocky body and an interdorsal ridge with no other conspicuous markings. The first dorsal fin is triangular and taller than the dusky, which is about all you have to go on when you try to make an identification of this free-swimming shark. They normally hunt in packs on squid and small fish bait balls.

Although the sandbar shark is abundant inshore and offshore from intertidal water to depths of 280 metres, I have witnessed the sandbar feeding behind the breakers at sandy beaches and surf zones, feeding on sardine and squid shoals. Off the islands it is concentrated near the edge of the banks. Although it may reach 230 centimetres in length, more distant specimens in Mauritius and the Seychelles are usually less than two metres.

In some parts of the world the females tend to stay near the shore, while the males seem to prefer the cooler offshore waters. The same may be true of the waters off Mauritius.

One specimen of *Carcharhinus plumbeus* was recaptured ten years after it had been tagged and released; it had grown only 50.8 centimetres. It is thought that the sandbar shark has a lifespan of approximately twenty years, but experts disagree about the length of time that it takes for it to reach sexual maturity as early researchers believed they were able to breed after two years, but it is now thought that it takes over thirteen years.

Solitary individuals tend to be cautious, but when moving in a pack they are unafraid of divers. The shark has triangular serrated upper teeth and feeds on a wide variety of fish, octopus, shrimp and crabs. Despite its formidable dentition there is no record of an attack on man.

SICKLEFIN LEMON SHARK
Negaprion acutidens
This shark inhabits the warm tropical waters of the Indian Ocean. It can be found off the coast of Mozambique, Madagascar, and the other Indian Ocean islands.

Divers can observe the sicklefin lemon shark over the coral reef or on the seabed, swimming slowly or resting on the sand. This shark,

Wolfgang Leander

often an inhabitant of shallow sandy areas in lagoons and estuary mouths, may also be found at outer reef shelves to depths of 30 metres. Juveniles are often seen over shallow reef flats with the tips of the dorsal fins breaking the surface.

This shark is pale yellowish brown, with a large stocky body and a short snout, and its dorsal fins are nearly equal in size. Although its teeth are smooth-cusped, some sharks over a metre and a half in length have slight serrations on the teeth of the upper jaw.

These sharks mature at a length of two metres and can reach a size of just over three metres. They feed on bottom-dwelling fish, and rays. When divers enter the water they usually disappear quickly, but like all sharks, they react aggressively when molested, and there are records of retaliation attacks on divers.

WHITETIP REEF SHARK

Triaenodon obesus
The whitetip reef shark is found from north KwaZulu-Natal, South Africa, and then northwards up the East African coast and across the Indian Ocean islands to include the Maldives. They are regarded as territorial sharks, spending most of their life around the same reef areas.

This reef shark is rarely longer than about 160 centimetres in length, a slender grey shark with very noticeable white tips on its first dorsal and upper caudal fins, sometimes even on the second dorsal fin tip. It has a very short broad snout with prominent brow ridges that give it a sardonic look. The nostrils have distinct tubular nasal flaps. The colouration of the upper body is grey to copper brown and a white underbelly.

Normally found in clear shallow water on reef flats, this predator is often also seen in deep water and you may encounter it while diving in crevices, caves, under ledges and resting on the sandy bottom. It is essentially a 'bottom dweller', feeding on small fish, eels, octopus and crabs. It is more active after dusk.

The diet of larger sharks in this range often includes smaller whitetip reef sharks.

Divemasters operating in Mautirius have discovered the most popular resting sites of these sharks in areas such as Lobster Canyon on the east coast of Mauritius and although the sharks usually keep their distance, a few professionals have been able to feed certain individuals. Oscar, a female whitetip reef shark that visits Meridien, had taken a liking to veteran divemaster Mich Frederic and would readily take food from his hand. This takes experience, however: the sharks may become agitated when disturbed by divers and, like most fish, will dart off into the distance.

SILVERTIP SHARK
Carcharhinus albimarginatus
The silvertip shark is found around all the Indian Ocean islands and down the east coast of Africa and as far south as KwaZulu-Natal, South Africa.

Silvertip sharks are large, may reach a length of three metres and are easily recognisable. The white tips of both upper and lower lobes of the caudal fin, the white tips on the first dorsal fin and the thin pectoral fins are conspicuous field marks. This shark has an interdorsal ridge between the dorsal fins and a white band running along the flanks. They normally have a coppery dark grey upper body and white underbelly. The silvertip has good dentition for eating all sorts of fish and has no problem ripping off chunks from dead mammals.

They are both coastal and pelagic. The juveniles are usually found inshore and the mature ones in deeper water. These sharks are seen in deep passages leading to coral lagoons, reef drop-offs around islands or deep ocean banks.

Divers will either encounter a single shark or a pack. The mature silvertip is a bold and inquisitive shark and it is not unusual for them to swim directly up to divers. This direct approach can be a bit daunting to divers the first time they encounter this behaviour. They may even swim alongside with a pectoral fin almost touching you.

This was the first-hand experience of Marilou McKay diving with Jeff, at the drop-off at the island of Karedu in the Maldives. Although, Jeff reckons the attraction was the green metallic shine of her stinger suit. The silvertip can get quite aggressive especially when competing for food and is notorious for the classic display of body language when threatened.

OCEANIC WHITETIP SHARK
Carcharhinus longimanus

The oceanic whitetip shark, as its name states, is basically a deep ocean shark. However, they do venture inshore, especially when the continental shelf is closer inshore. I have come across these sharks near the deep-sea trenches of the north KwaZulu-Natal coastline and the Comores Islands. They are found from the South African coast all the way up to Sri Lanka, which includes the Indian Ocean islands.

This large shark, with its long, rounded, white-tipped pectoral fins, is unmistakeable! You may see it swimming in a leisurely fashion at or near the surface with its huge paddle-shaped pectoral fins spread out. This shark is deceptive, it can move fast when aroused and it is very aggressive, especially when hunting for food. Divers report that it can be both bold and persistent; it will approach divers repeatedly despite attempts to fend off their investigations. Many attacks on shipwrecked people have been attributed to the oceanic whitetip.

With its large, broad, strongly serrated teeth the oceanic whitetip shark can eat virtually anything that it can catch. It feeds on dead whales, pelagic game fish, seabirds, bony fishes such as sardines, and cephalopods and turtles. Game fishermen off the islands report that it will take a marlin bait and will also devour garbage and carcasses.

Although oceanic whitetip sharks may be found cruising near the surface or in water as shallow as 40 metres, particularly off oceanic islands or in areas where the continental shelf narrows, they generally prefer deeper water. They congregate in large numbers around a food source. They segregate according to size and sex.

The shark, born with a length of 65 centimetres, matures when it reaches about 1.7 metres, and it may reach a length of three metres. The upper body is grey-bronze and it has a white belly. There is white mottling on the tip of its high, rounded first dorsal fin and at the ends of its long pectoral fins.

ZAMBEZI SHARK

Carcharhinus leucas

The zambezi shark inhabits the Indian Ocean from South Africa up the entire East African coastline and the western coast of Madagascar. The zambezi shark patrols the inshore reefs from sixty metres depth to the shallows of the inshore channels. They are observed in lagoons, upstream of rivers and harbours. They are particularly fond of preying in muddy river water and have a good tolerance to fresh water: ergo – do not dive or swim at river mouths or in muddy waters after summer floods.

They prefer the warmer subtropical and tropical sea temperatures although some of the rivers like Port St Johns get to 16 °C, which has seen several shark attacks of late, all categorised as being from zambezi sharks.

This shark is prone to parasite infestations and it is thought that the change in water temperature and salinity may help as a cure.

The 'zambie' is a large robust-bodied shark with a blunt snout. It has triangular cutting teeth in the upper jaw and a wide mouth. It has no interdorsal ridge. They are grey to coppery brown on the upper body and have a white underbelly. This shark can attain a length of over three metres and has a wide varied diet consisting of skates, sand sharks, turtles, rays, smaller sharks and bony fishes.

It is also a known scavenger in muddy waters and to this end, it has caused havoc with bathers in these conditions. A characteristic of a zambezi attack in poor water visibility, is the bump…. then bite! This is known as 'mouthing' in shark parlance.

This shark is territorial and possessive of the reef that it patrols at the time, it does not hesitate to investigate any intruders and occasionally are seen in pairs, but mostly a single shark is sighted. They normally arrive with their armada of remoras.

The zambezi is ranked the number three for shark attack. Some spectacular sightings have been had on the Protea Banks on the lower South Coast and deeper reefs of the upper South Coast of KwaZulu-Natal.

GREAT HAMMERHEAD

Sphyrna mokarran

The great hammerhead is found all along the coast of Africa and the Indian Ocean islands. Fishermen and divers often encounter them on their single nomadic journey in search of prey. They are easily spotted at the surface with just the dorsal and upper caudal fin exposed. Divers have encountered them on deep drop-offs when drifting on a wall dive. Fishermen trawling for billfish generally find them taking their baits. Common places for reported sightings are Mozambique, Ponta d'Ouro, Xai Xai, Inhaca Island and Pemba. In Mauritius they are found near Flat Island, Round Island and Gunner's Quoin and in South Africa from the Transkei coast northwards all along the KwaZulu-Natal coast.

It is a large shark, reaching a length of more than 3.5 metres. It has a large, flat hammer-shaped head, a dark grey or brown back and a white belly, with a high sickle-shaped first dorsal fin. It is the largest of the hammerheads; they range about 60 centimetres at birth and mature when they reach about two metres.

They are nomads, ranging from the surface to the seabed, and feed on a whole variety of bony fishes and squid. A very manoeuvrable and swift swimmer, its hammerhead acts like a hydrofoil. It uses its unique head to eat from the seabed as well.

This species is considered dangerous, though few attacks have been confirmed. It is very inquisitive and is known to approach divers, especially spearfishermen, and this may be interpreted as aggressive behaviour, particularly if the shark is large. Always treat a great hammerhead shark with respect with its strong serrated teeth − it is capable of inflicting severe injuries.

SCALLOPED HAMMERHEAD

Sphyrna leweni

The scalloped hammerhead moves around the Indian Ocean and is observed around the islands, the coast of South Africa and up northwards up the East African coast. It prefers swimming along

Wolfgang Leander

reef drop-offs and cliff faces. It swims between deep and shallow water. Usually they are seen in large migratory packs, consisting of all sizes. The loners that are encountered are mature sharks. Aliwal Shoal on the upper South Coast of KwaZulu-Natal is a popular place to sight these migratory packs of scalloped hammerheads, especially during the months of July to October.

The scalloped hammerhead has a curved hammer with a central notch. It has smaller dorsal fins compared to the great hammerhead shark. This shark grows to a length of around three metres. The dorsal surface is a coppery brown to grey and the underbelly is white. Its teeth are smooth edged.

The scalloped hammerhead mainly feeds on small bony fishes and squid. They are usually attracted to the divers' bubbles as they expand and rise to the surface. So observing them circling your bubbles is a known behaviour.

They are not regarded as dangerous, but do get excited about speared fish and will get brave for an opportunistic feed.

BLUE SHARK
Prionace glauca
Blue sharks are found in the deeper waters of the Indian Ocean's tropical latitudes. Fishermen trawling for sailfish often hook the blue shark.

This shark really is blue – bright blue as the name describes it accurately. It has a dark-blue back, brilliant blue flanks and a white belly. It is unquestionably one of the most beautiful and graceful of sharks. With its large slender body, long pointed snout and long, thin pectoral fins it moves with the elegance of a jet fighter.

The body shape and small lateral keels in front of the caudal fin identify the shark as a surface-feeding fish and it preys primarily on small sharks, shoaling fish and squid. It also feeds on seabirds and scavenges on marine mammals. It may venture close inshore at night.

Divers usually see this shark cruising, pectoral fins spreading as it glides, and its dorsal fin cutting the surface. Unafraid, it does not hesitate to circle or approach the diver. It often becomes bold when food is available and spearfishermen consider it a menace. It has the size, dentition and disposition to inflict fatal damage.

There is a specimen on display at the Natural History Museum in Port Louis; an International Game Fishing Association (IGFA), and a world-record blue shark of 400 pounds (183.6 kilograms) was landed in Mauritius.

MAKO SHARK
Isurus oxyrinchus
The mako shark is found all over the Indian Ocean and has been caught as far south as the Agulhas Banks off Struisbaai, in the southern Cape coast of South Africa. It is predominantly a deep-water shark and often found close to land where the ocean floor drops away, such as the Indian Ocean islands.

The diet of the mako shark mainly consists of bony fishes, smaller shark and squid. This shark relies on speed and manoeuvrability to catch the game fish and squid.

Game fishermen may see a mako shark with its dorsal fin slicing the surface at great speed as it scans the water for a meal. The mako shark is known to be capable of leaping six metres out of the sea, in pursuit of its prey.

Divers have observed these sharks hanging around the edge of the water visibility and slightly below when decompressing, at in water decompression stops. My most memorable meeting of a mako was on decompression at the Coopers Light Wreck off the Durban Bluff, KwaZulu-Natal. It is worth noting that this shark has been recorded in several fatal attacks.

Although it is a close relative of the great white shark, the mako has the non-serrated awl-shaped teeth of a fish-eating shark. This shark has a long pointed snout, large black eyes and a torpedo-streamlined body, pronounced keel just forward of the lunate tail. It can only be regarded as one of the magnificent species of any ocean. Mako sharks are known to reach lengths of three metre and more, the females tend to be bigger.

TIGER SHARK
Galeocerdo cuvier
The tiger shark is found all over the Indian Ocean from the tropical to subtropical waters. Regardless of the tales that this is only a deep-water reef shark, this shark is found in shallow sandy island lagoons, inlets and shallow reefs just behind the surf zones.

The tiger shark has a large blunt head shape, large black eye, and slender body. Like the mako, this shark also has the lateral keel before the tail fin. The upper body is coloured anywhere between light to dark brown, with darker stripes running down its flank; these fade with age. The underbelly is white to sometimes yellowish.

Its appetite, alone, makes the tiger shark dangerous, it will eat sharks, rays, turtles, seals, dugong, dolphin, birds and anything that it can get between its jaws. The KZN Sharks Board discovered

Wolfgang Leander

the remains of a human skull and jawbone in the stomach contents of a tiger shark caught off Leven Point, northern KwaZulu-Natal. In another dissection, the hubcap of a car was removed from the stomach and more than one underwater photographer has battled to retain his camera after it had been 'chomped' by a tiger ...

Its cockscomb-shaped serrated teeth can slice through bone and flesh like a guillotine. Fortunately for divers, most Indian Ocean waters are rich in food, and attacks on human beings are not commonplace. A foolish practice of spearfishermen lately is to chum the area to attract fish, this is most unwise in known shark waters. A tiger shark took the life of a spearfisherman doing this off the northern KwaZulu-Natal coast.

Remember where there is fish activity, there will inevitably be sharks. The tiger shark is a skilled hunter and is ranked the number two threat for attacks on divers next to the great white shark.

Its length at birth is approximately 60 centimetres and it matures when it reaches just over two metres. It may eventually reach five and a half metres or more in length. The author of this book, Al Venter, recalls that while diving in the Cape Verde Islands of the Atlantic with former East London salvage expert Peter Sachs, a monster of six and a half metres was landed by one of the islanders some years ago.

GREAT WHITE SHARK
Carcharodon carcharias

The great white shark is mainly found in the temperate waters to subtropical waters, more frequently in the winter months. They are pelagic but spend periods feeding in coastal waters and around offshore islands where there is an abundance of food. This is very much seasonal, with bony fishes and mammals.

The southern Cape Coast of South Africa is known for its greater concentration of great white shark, especially off the islands inhabited with seals and marine birds. Divers have also encountered them in the shallows just off the kelp beds. Great white captures have been recorded in the near zero visibility of the Richards Bay shark nets. This proves that location or water visibility is no guarantee that you may not encounter this species. Fossilised teeth of the extinct Megladon species have also been excavated in the Richards Bay area.

The unfortunately 'notorious' great white shark is a rare visitor for the diver. Adults prefer to hunt marine mammals of the cooler seas, but have also been encountered feeding off carrion in the warmer tropical waters of the Indian Ocean. Where there is a dead whale off the southern African coastline, you will find great whites feeding, usually vociferously.

The great white shark has a powerful torpedo-shaped body, robust conical-shaped snout, large triangular teeth, which are heavily serrated, and black eyes. It displays large pectoral fins, a prominent keel on the caudal peduncle and a lunate tail. The upper body colour will vary from pale grey to black and the underbelly is white. The tips of the underside of the pectoral fins will have black blotches marking them.

The diet of the great white is made up of a variety of sea mammals, shark, bony fishes and squid.

Although a juvenile great white shark feeds primarily on fish, it has been known to attack human beings, and it should be treated with caution at all times. The great white shark is regarded as the number one threat to divers and should always be treated in this respect. The chance of even the most active scuba diver ever seeing a great white shark in tropical waters is remote.

In the subtropical waters of the northern KwaZulu-Natal, the owner and operators of 'Mokorran Dive Charters', Daryl, Clive and Debbie Smith have had some magnificent encounters off Rock Tail Bay. I recall a sighting at Two Mile Reef at Sodwana Bay by Rod Hastier, an ex-KZN Sharks Board diver. In the temperate waters of the Cape, André Hartman, the famous 'shark diver' who we have seen often enough on *Discovery Channel* is another diver and operator that has had numerous encounters in and out of the shark cage.

On one occasion that I vividly recall, Al Venter, Marie Levine of SRI and I had an enormous great white break the surface near our dive boat, just moments after ending a wreck dive in the southern Cape. Al was the last to board and, as he left the water, the shark popped its huge head out of the water alongside him and viewed him carefully with his big black eye before slipping back into the sea. That was a really hairy experience: water visibility was barely a metre and chances are that this predator had been there throughout our dive …

If you ever have the chance to do a cage dive with a reputable operator to observe a great white shark underwater, take it up! It will be the most amazing and humbling experience, it's up there with the best that life has to offer.

WHALE SHARK
Rhincodon typus
The whale shark is found in the tropical waters of the Indian Ocean and ventures into the subtropical waters during the summer months. Living in the open sea and venturing close inshore for food, the whale shark is encountered around the tropical Indian Ocean islands and the East African coastline southwards to KwaZulu-Natal, South Africa.

The whale shark is aptly named, the largest living fish growing to an age of approximately 70 years and lengths of up to 16 metres.

It has a wide mouth situated at the beginning of the broad blunt snout. A small eye is located each side, at the forward part of the wide flat head. A spiracle is located just behind each eye. The whale shark's body has three distinct ridges running along its length on each side, with a rough abrasive skin, with light spots and stripes on a dark grey to brown dorsal surface. The underbelly is white. It matures with a lunate tail with a lateral keel forward of the tail fin. As a predominantly filter feeder, it is equipped with ten filter pads and over 300 rows of tiny teeth.

The whale shark feeds on plankton, krill, small squid and small shoal fish.

I had an amazing experience with Rob Allen one day travelling back from Sodwana Bay to Cape Vidal on his boat, when we came across a bait ball of small fish being preyed upon by shark off Leven

Point. In amongst all this activity was a single whale shark. I had Al's underwater camera and jumped into the sea to get some footage. I swam up close to the whale shark's back, which was hanging vertical in the water, feeding. This gave my back protection from the other sharks that were a bit too many and too close.

Meantime, there were shoals of sardines swarming the whale shark for protection from the other predators and, using this advantage, the whale shark was sucking in huge volumes of fish through its cavernous mouth. This went on for a while, and then quite abruptly, the activity stopped. No more blacktip and long nose blackfin sharks! It all became clear when a huge tiger shark with remora's trailing, passed beneath us.

Divers may experience these sharks during the summer months along the coastline and surrounding offshore reefs. During a flight up the Zululand coast from Cape Vidal to Sodwana Bay, a young spearfisherman, Warwick Meyer and I counted 25 whale sharks swimming up the coast beyond the breakers during December one year.

A whale shark is a living, moving reef. Swarms of minute para-sites graze over the ponderous hulk. Often a prodigal son, dorado or kingfish are seen swimming with them. Divers who have encoun-tered whale sharks have hitched rides. They have clung to its lips or the huge first dorsal fin or pectoral fins and have been taken on an unforgettable ride through the sea until the shark tired of the game and sounded.

Rob Allen, Geremy Cliff and I tagged the first whale shark off Cape Vidal in the late 1980s. This was the first step to monitor their movements along the African Coast.

BASIC SHARK LORE FOR THE DIVER

Sharks are magnificent creatures, aesthetically beautiful creatures, moving through the sea with incomparable elegance.

It is the diver's responsibility to enter their domain with knowl-edge, respect and personal safety. I strongly urge divers to further their sport or profession and attend a course provided by shark-experienced course providers. One such person would be Debbie Smith of 'Diving with Sharks'.

Of the more than 300 shark species known to exist, only a few are potentially dangerous to man. A single attack has been recorded off Mauritius where a small shark lunged after a spearfisherman's catch and nipped his arm.

The South African and Mozambique coastlines have a different history compared to Mauritius — which only proves that divers can make an extra effort to reduce fatal encounters.

There are some elementary guidelines to follow when diving with sharks:

1. Never harass or molest a shark in any way.

2. Some sharks are not only hunters but scavengers as well; so don't dive where fishermen are cleaning their catches or around seal colonies when they are giving birth to young pups. It is inadvisable to swim with an open wound for the same reason. Avoid water with poor visibility as some sharks often feed in turbid water near river mouths.

3. Always dive with a buddy and stay close together, especially while diving at night as most species of shark hunt their prey after dusk.

4. If sharks become too nosey and bold, consider leaving the water. If necessary, keep your back against the rocks or go to the surface, back to back, with your buddy. If you are really under attack, remember that a shark's snout is very sensitive. In a life-threatening situation, a stiff blow on the snout may work. However, there are cases in which such action has made the shark more aggressive. That said, this author would much rather have something in his hands to fend off an attack by an aggressive shark. (A camera and a speargun have been used with success.)

5. Don't move erratically and noisily, especially when there are sharks about. Sharks are stimulated or irritated by excessive movements, causing vibration through the water and will be picked up by their sensory systems. Instead, take a few easy breaths, move slowly and blend into the environment and enjoy the experience of seeing these wild magnificent creatures.

CHAPTER ELEVEN

WHALE SHARKS:
A 'MONSTER' OF AN EXPERIENCE

Marine research scientist Morgan Riley has made a career of the close-quarter study of whale sharks. As the founder-director of the Maldives Whale Shark Research Programme, he recently returned to Britain from the Maldives, where he led a series of comprehensive research initiatives.

O f all the experiences that the ocean can conjure from its depths, a close encounter with its largest shark, indeed the largest fish yet discovered by science, is surely among the most awe-inspiring. As the cavernous maw, large enough to accommodate your average diver with room to spare, emerges like an apparition from the blue, you cannot can't help but gulp.

The mouth may be more than a metre and a half across and is lined with row upon row of 6,000 (albeit tiny) serrated teeth. Despite your higher conscious knowing that you are not an item on this great creature's menu – as each of these teeth is propelled towards you at deceptive speed by a body and tail the length of an average city bus – your primeval brain cannot help but question whether its disproportionately diminutive counterpart behind the gigantic jaw is aware of this. Yet, with about a metre to spare, the mouth closes, its head rolls to one side and the whale shark vacantly gazes straight into your eyes.

Though whales are often bigger – they are warm-blooded mammals – it is the whale shark, scientifically classified as *Rhincodon typus*, which is the largest known fish on the planet. Just how large whale sharks can become is open to dispute. Invariably they are anecdotally reported as being larger than the boat on which the enthusiastic observer finds himself, regardless of the vessel's size.

In reality, the vast majority of whale sharks recorded by specialist research teams dotted throughout the tropics are between four and eight metres long. Historical fisheries records document lengths of up to 13 metres and estimates of whale sharks up to 15 metres (from experienced divers in the eastern Pacific) seem plausible. The largest creature in this category that this author has observed was a paltry ten and a half metres. Yet, despite their size and an abundance of myth, very little is actually known about whale sharks.

This chapter will attempt to debunk some of the myths and unveil some startling truths.

As early as 1786, one ship's log described a whale shark as twelve metres long and three metres wide. The description went on that '... [it was] beautifully spotted like a leopard; in some parts, the spots on its body resemble that on the peacock's tail.' However, it was not until 1828, when Andrew Smith, an English physician, described the whale shark from a 4.6-metre specimen harpooned in Table Bay, South Africa, that the species was formally recognised by science.

In spite of its widespread presence in the oceans of the world, the ability to encounter a whale shark in the wild has, until fairly recently, been regarded as an exceptionally lucky experience. Indeed, despite its size and striking pigmentation, there were only 300 encounters with these creatures documented by the mid-1980s. Since then, cheap travel, water sports and waterproof cameras have contributed towards more than 15,000 recorded encounters with the species, with the number increasing each year.

The scientific community has learned to take into account factors such as lunar and tidal cycles, weather conditions, the state of the sea as well as the amount of maritime traffic to predict the presence (or absence) of whale sharks at specific times at some near-shore locations with a degree of accuracy that ranges from hours to months, although, as with all encounters with wild animals, luck remains a significant factor.

Sporadic reports of offshore aggregations of more than a hundred whale sharks currently seem unpredictable, but with the necessary resources, dedication and careful analysis, scientists may one day be able to predict these as well. However, while one might be excused for thinking that with the abundance of data available to researchers, the intimate secrets of whale sharks might now be known to science, the truth is that many aspects of the life of the whale shark remains hidden.

There is an old anecdote about a passer-by asking a man searching on the floor under a street light what he has lost. 'A contact lens' is the response. 'Where did you lose it?' asks the passer-by. 'Over there,' the man says, gesturing a little way off. 'So why look here?' asks the bemused passerby. 'Because,' answered the searcher, 'here the light is better.' To some extent this anecdote can be applied to observations of whale sharks.

The vast majority of the 15,000 recorded sightings of whale sharks come from a handful of relatively small, seemingly isolated areas where some degree of predictability in the presence of these delightful creatures has caused a growth in ecotourism. These locations include Mexico, Mozambique, the Maldives, the Philippines, the Seychelles and Western Australia. Consequently, it is generally accepted that whale sharks are largely a tropical or subtropical species even if some are occasionally encountered in colder climes. Yet, one must consider the circumstances necessary for an encounter with a whale shark to be recorded.

First, someone must be present. Second, they must be able to see the shark. Third, they must believe that recording its presence has some value. And finally, they must have the means of transmitting that recording to some form of database.

When in or on the sea, people are most likely to be within a few hundred metres of the coastline with a relatively warm climate where there is an established tourism infrastructure. To see the shark, the surface conditions must be relatively calm, the visibility good and it must be daylight. To think that recording the presence of a shark has some value depends on perception, personal interest and culture of the observer. To transmit such a recording requires access to technology as well as knowledge that such databases exist.

Consequently, it is of little surprise that those locations listed above, (where these criteria are met, and where established whale shark research groups promote the recording and submission of reports to a database) provide the overwhelming majority of whale shark recordings. However, this reflects human behavioural preferences, not necessarily those of the sharks, and should be borne in mind when drawing conclusions about whale shark behaviour and ecology.

Of course, the sharks do have to be there and it is axiomatic that the specimens encountered have proved to be tolerant of the presence of humans as, unlike whales, whale sharks do not need to broach in order to breathe.

Currently, whale shark databases tend overwhelmingly to record juvenile, and usually male, specimens. The total number of recorded sightings of whale sharks smaller than three metres might be counted on your fingers and toes while mature adults – generally classed as those of over nine metres – represent only a tiny fraction of the documented individuals. This, of course, does not mean that the whale shark global population contains very few adults or young sharks. Instead it suggests that most whale sharks spend the major part of their existence in habitats that remain hidden from human observers, or at least from humans who are willing and able to record them.

Data retrieved from recording tags attached to whale sharks support this conclusion. They indicate that these creatures are able to tolerate water temperatures below 4 °Celsius while diving to depths that exceed the 1,400 metre pressure tolerance of the tags and can travel thousands of miles in the process.

Furthermore, these tagged animals spend most of their time at depths where they would be undetectable to humans and, when they do approach the surface, it is usually at night. It is also important to bear in mind that a shark that has been tagged must have been inclined to venture, at least once, into a habitat where a scientist was ready and able to attach said tag.

Their less amenable brethren may be found to exist at even greater depths, further offshore and completely hidden from current research projects for the duration of their lives.

Just as studies of whales often use the scarring and colouration of tail flukes to identify individuals, whale sharks are identified by what is termed their 'spot patterns'. Typically, whale shark photo identification employs images of the flanks of these creatures, concentrating on an area immediately above the pectoral fins.

Research teams augment their databases by urging tourists to send in their photos of whale sharks and to provide dates, locations, sexes of the animals and estimates of length. These are then compared to existing whale shark image libraries. In this manner, the animals' locations can be non-invasively tracked and at minimal expense.

These databases have subsequently been used to estimate population trends, determine sex and size ratios, and, by comparing injuries, used to identify potential threats to the species. Research teams are also in the process of developing methods to precisely record the length of each shark during such encounters, most recently using a system of lasers. In this way, rates of growth in wild sharks can be calculated, providing valuable data for developing conservation and management plans for the species.

However, other than only representing a non-representative fragment of the total global whale shark population for the reasons above, this information has another weakness. To identify and monitor individual sharks by using their spot patterns, the assumption is made that each spot pattern is unique and that the pattern does not change.

The first is probably true, although as image libraries grow in size, it becomes increasingly likely that a new shark's spot pattern will resemble that of an existing shark in the database sufficiently for cases of mistaken identity to take place. Still, such errors can be minimised by using images of both sides of the same shark and by the rigorous cross-checking of data.

The assumption that a spot pattern does not change is almost certainly false in the long term, because whale sharks' bodies grow disproportionately. If a shark is regularly observed, the image library can be updated and slow incremental changes will not cause a problem. But if a shark is not seen for a decade or more, its spot pattern may have changed beyond recognition the next time it is photographed.

Also, any kind of dependence on wounds, other than partial amputations to aid identification, is unlikely to be reliable because we are now aware that the animals have the ability to heal at a phenomenal rate.

It is clear then that we as scientists have a problem: through its behaviour and biology, the whale shark has almost conspired to hide its secrets from those of us involved in research. And while experts can be eager to draw conclusions and to make bold statements to justify their status and research grants, the honest answer more often than not is that we simply do not know. We don't even really know how many whale sharks there are in our oceans.

Estimates from genetic studies range from 25,000 to 500,000 breeding adults, but even these figures are speculative, which is why working with whale sharks is both exciting and demanding: the very next discovery might turn existing theories totally on their head.

For example, common knowledge, and until recently your author, would tell you that whale sharks eat plankton, occasionally supplemented by small fish and squid. Like their filter-feeding cousins — basking sharks and the appropriately named megamouth sharks — whale sharks cruise the oceans, mouth agape which allows them to filter large volumes of water over their specially adapted gills where prey is collected for ingestion.

However, the filter screens of the whale shark are denser than those of the basking shark and the megamouth shark, making for a filter apparatus that is a good deal more efficient for short suction

Sijmon de Waal

intakes (in contrast to the flow-through system described above and favoured by the other two species). This ability greatly expands the range of feeding opportunities available for whale sharks, including recent observations of whale sharks being spotted sucking decomposing flesh from the cadaver of a whale and stealing a bag of chum off the swim platform of a dive boat.

In New Guinea waters, whale sharks tend to congregate daily to feed on the by-catch routinely discarded by local fishermen and regularly try to suck fish from arrays of storage nets suspended beneath fishing platforms. While this hardly makes whale sharks veracious predators, it does suggest that they are adaptable and capable of opportunistic behaviour and should therefore not be regarded as being exclusively filter-feeding planktivores.

With so few records available of neonatal or adult whale sharks – coupled to a relatively porous historical record – estimates of almost all the parameters of the species' life cycle must be considered merely as indications. For what it is worth, it is widely speculated that whale sharks mature when they are between eight and nine metres long, and perhaps 30 years of age. The accepted consensus is that they may have a lifespan of between 60 and 100 years.

As with most other shark species, whale shark mating has never been observed. On the basis of a single discovery of a live and almost fully developed embryonic whale shark in an egg case trawled in the Gulf of Mexico in the 1950s, it was believed for decades that whale sharks were oviparous (egg-laying), like the majority of Orectolobioformes, the group of sharks in which they are categorised.

However, a 10.6 metre, 16-ton gravid female whale shark captured in 1995 off the east coast of Taiwan, suggests otherwise, and that the species is actually oviviparous, retaining the neonates within the uteri to allow further development (as in nurse sharks which are a close relative). This female contained 297 embryos, evenly split between male and female, the largest litter size reported for any species of shark to date.

The prenatal sharks could be separated into three distinct developmental stages, suggesting that successive batches of pups are born over several months. One also needs to be circumspect about drawing firm conclusions from a single animal. After all, it is possible that this litter was not representative of the species.

Of the embryos, more than half were 58 to 64 centimetres in length, and were free-swimming, without a yolksac. Although the duration of whale shark gestation is unknown, these were probably ready to be birthed as neonate specimens occasionally caught in fishing nets and that have ranged from 55 to 67 centimetres in length.

Two of the whale shark pups taken from the Taiwanese gravid female were aquarium-reared, providing significant post-natal growth data. One pup, originally 60 centimetres long and weighing a single kilogram, grew to 1.39 metres and 20.4 kilograms over a period of four months before it died of septicaemia. The other pup grew from 60 centimetres to almost four metres in just over three years. Those figures alone indicate prodigious growth capability.

Usually a large litter size and rapid neonatal growth rate suggest a high level of neonatal mortality, borne out by the fact that immature whale sharks have been found in the stomachs of blue sharks and blue marlin. Once a whale shark reaches adulthood, its enormity and a dermal layer that is as much as 14 centimetres thick – further protected by enameloid crowned scales – provide perhaps their greatest defence against predation.

In spite of this, a fatal attack on an eight-metre whale shark by two orcas has been documented and fresh wounds provide graphic evidence that whale sharks are occasionally subject to attacks by large predatory sharks. However, as with most, if not all shark species, the main threat to the whale shark community comes from us humans.

Due to the large size and docility of whale sharks, the species is an easy target. Unfortunately, innate characteristics, such as large size, slow development and late maturation, delayed and infrequent reproduction as well as long lifespan, almost certainly mean the species has a very low exploitation threshold. Indeed, historical artisan fisheries catching relatively low numbers of whale sharks for liver-oil to treat the hulls of wooden boats caused serious declines in local abundances, a trend that suggests that even small traditional fisheries may ultimately be unsustainable.

Tragically, from the early 1980s, the demand for whale shark products escalated exponentially, fuelled in large part by the demand for fins and meat in both Taiwan and mainland China. The Taiwanese fishery rapidly exhausted their own apparent domestic whale shark stock, resulting in increased demand for imports. More than 1,000 whale sharks were taken by the Indian whale shark fishery in 1998, dropping to 279 in 1999, then, despite increased fishing effort, to only 160 in the year 2000 suggesting a wholly predictable crash in whale shark numbers.

Fortunately hunting is now prohibited by domestic legislation throughout much of the species' known range as some (but not all) individual governments realise the value of the living sharks to their ecosystems and local economies.

The Convention on International Trade in Endangered Species, along with certain United Nations agreements do provide some limited international protection, but it is an extremely difficult task to police all the oceans of the world for transgressors, and incidents of whale sharks being deliberately targeted continue.

This kind of opportunistic illegal fishing activity ultimately represents the single greatest immediate threat to the survival of the species.

What is clear is that current unilateral conservation measures in any single political jurisdiction are inadequate.

Management plans for the global protection of whale sharks require ocean-basin-wide, if not global, cooperation. Evidence for this comes from a dramatic 96 per cent decrease in whale shark sightings in the period 1991 to 2001 in the Andaman Sea, and a reported 40 per cent reduction in whale shark abundance at Western Australian locations, despite protection under their respective national legislation suggesting unsustainable mortality in other parts of their range.

However, such evidence is disputed, due in large part to the difficulty in tracking population trends in an elusive species. It is possible that the creatures might have moved elsewhere or simply stopped approaching the surface.

The accidental effects of human activities may also threaten whale sharks. The deterioration and destruction of important habitats could eventually be a serious threat to whale shark survivorship, as could also be rapid climate change.

El Niño Southern Oscillation events occur naturally every three to seven years, through the interaction between the ocean and atmosphere in the tropical Pacific. But they are also sensitive to background climate change and have increased in frequency and strength during the past half-century. When the El Niño cycle is particularly strong, the changed weather patterns have profound ecological consequences, including an apparent effect on whale sharks – although in such a complex system it is almost impossible to determine direct causation.

A further human cause of whale shark fatalities is collisions with ships – a problem which is probably increasing as the volume of international shipping grows and ocean-going ships become ever larger and faster. At several research locations more than half of the whale sharks displayed some evidence of boat strikes.

As mentioned above, the massive increase in recorded human-whale shark encounters is partially due to a huge increase in recreational diving and boating activities, especially the current vogue for ecotourism. Minor abrasions or lacerations characteristic of small-boat propeller strikes have been noted on whale sharks at most, if not all, of the sites where an industry has been developed for whale shark ecotourism, and in at least some cases are caused by the operator's vessels.

When exposed to large numbers of people in the water, whale sharks often dive precipitously to deeper water. Importantly, despite the declarations of primitive dive guidebooks and certain unscrupulous operators, whale sharks will not tolerate being touched. To do so is to display a selfish disregard both for the animals and fellow dive guests.

Conversely behavioural studies show that with each additional metre that swimmers stay away from the shark, the less likely the whale shark is to react to their presence.

An area where human-whale shark encounters are poorly managed may be avoided by the species in future, possibly excluding them from important habitats and destroying the industry dependent on their presence. With a reduction in the perceived value of whale sharks as a living resource to the local community they may again become targeted by fishermen. Realising these risks, many nations now regulate the industry to improve the management of whale shark encounters.

Typically, there are guidelines in place in most (but not all) tourist destinations that prohibit the touching of these creatures. Regulations with regard to flash photography and limiting the numbers of tourists to whom the sharks are exposed are also in place, as are laws that control boating activity.

Hopefully all these measures will have a positive effect on whale shark populations for the future. If not, they may become more evasive and future generations will revert to the seldom-seen phenomena of previous centuries.

Sean Botha

CHAPTER TWELVE

PROTEA BANKS:
AN EXCITING SHARK DIVE

Pugnacious and, as someone said – 'front-end loaded' – the bull or zambezi shark has always been a feature of diving at Protea Banks near Margate in South Africa. It can be a demanding dive and is not an experience for everybody …

*D*ivesite is arguably the most prominent dive magazine in the country: glossy and in glorious full colour, it appears quarterly as a conventional magazine and, concurrently, on-line at www.thedivesite.co.za. Its editorial content excels in keeping underwater enthusiasts appraised of what is going on in the dive world and its editor, Murray Jackson, is not afraid to 'mix it' when there are contentious issues that need to be tackled.

The publication makes a point of highlighting shark diving experiences, wherever these might be encountered: a recent report threw the spotlight on Protea Banks on the South Coast. The report, both instructive and candid, came from veteran shark diver Fiona McIntosh.

She told her readers that 'if you want to dive with sharks, there are few places in the world that compares with Protea Banks, off Shelley Beach on the KwaZulu-Natal coast.' She goes on about the venue, some seven kilometres offshore and swept by currents, telling us that it is home to a huge variety of sharks and other big pelagic species including zambezi, hammerhead, raggedtooth, dusky, thresher, tiger and blacktip sharks. There is also the occasional great white, mako and bronze whaler, large numbers of rays and game fish and, of course, the usual reef dwellers.

'There are two main areas – the Northern Pinnacles and the Southern Pinnacles. The former is best dived in the winter to spring months (June to November) when raggedtooth sharks congregate on the reef to mate, while the high chance of seeing zambezi sharks on the Southern Pinnacles occurs between October and May.' That means that this is the most popular site in the summer months.

'You can enjoy baited shark diving on Protea Banks with African Dive Adventures. This is a rare opportunity, as South Africa is one of the few countries that offer the experience. Using a baiting technique that closely resembles the shark's natural feeding habit, divers (as well as snorkellers and non-divers on the boat) can enjoy the thrill of interacting with sharks without a cage. Encounters with large tiger sharks are almost guaranteed from March until June, while zambezi sharks come to the bait year round. So if you're a photographer, here's your chance to get that award-winning shot.

'Although Protea Banks does not have the colourful reefs to match those of Aliwal Shoal, the sharks are not the only attraction. The recently discovered site of 'Playground', just north of there, for instance, also offers canyons, caves and spectacular rock formations, as well as schools of pelagic fish.'

Diving zambezi sharks on Protea Banks with Roland Mauz – who with his wife Beulah, runs African Dive Adventures – is always an unusual experience. He has dived the area almost 3,000 times and while he has had occasional 'words' with predators about what bait goes where, or perhaps a little territorial one-upmanship, he has never been scratched by a shark, let alone attacked by one.

It was different 30 years ago when divers thought sharks might have been more aggressive, an attitude forged by films like *Jaws*, sensationalistic reports about 'man-eaters' and, obviously, lack of experience. As he says, 'the predators have become accustomed to us in the water and we are doing things undreamed of before … call it common ground, if you will, but it certainly makes for some interesting things happening underwater.'

At the same time, Mauz is not into gimmicks. There are those operators that like to put sharks into some kind of catatonic immobility, turn them upside down or balance them on their snouts, but that's not for him or his dive operation. 'Obviously we now know that for some reason, it is possible to do all that … but to my mind, it is pathetic and totally unacceptable. My attitude is to have a healthy respect for the creature and, in turn, accept back some kind of deference or acceptance, even from the sharks with which we dive.'

Sharks, maintains Roland Mauz, are totally misunderstood. They are also misrepresented by the public at large. Even worse, these predators are misused by the media, largely to sensationalise their

Roland Mauz (right) and his resident divemaster Jean Pierre discuss the next underwater outing. That followed their return to Shelley Beach after a morning dive, in which the author's partner Caroline was deposited in the middle of a pack of 14 guitar sharks.

reports and inflate sales. 'And that all comes at a cost, because these beautiful creatures – and yes, they are truly beautiful – are defenceless.'

To date, he explains, his group has performed several hundred baited shark dives, most of the time with him in charge. 'Never once have we experienced any kind of aggressive behaviour ... how could this be possible if all the old stories were true?

'Today', he says, 'we are cheeky enough to tease a four-metre tiger shark with a chunk of tuna carcass, waving the piece of fish in front of its jaws, as if to say, come and get it. By doing what is seemingly a stupid thing, we know from the number of occasions that we have experienced it – more than 400 times – that the shark could easily take the whole fish, and if it wished to do so, me with it. And a four-metre tiger shark could do that in a single lunge. Yet, we know very well that it won't ... it is all part of the game.'

Things have changed in other directions as well, reckons this experienced divemaster. Very early on he would take veteran shark diver Andy Cobb into the water and 'Andy would arrive with a Bangstick, until I told him to chuck the stupid thing away … and that was in the early 2000s.

'Then we had Karen Tredger who, after a shark dive, cleared the almost metre and a half of solid gunwale and landed back on board the boat after she had experienced her first contact with zambezi sharks on Protea Banks in the early 1990s. But that same Karen today adores the experience of a baited tiger shark dive … she treasures a photo I took of her right behind one of these enormous predators, her hands raised to display a pair of OK dive signs.'

According to Mauz, there almost certainly are more sharks around today than there were before, at least in his location.

In fact, he says, there are seasons when the sharks seem to be taking over. But there are also times, and it is also seasonal, when there are no sharks on the reef at all. Nobody can explain it, he states … it remains one of nature's many mysteries.

'On 6 January 2012, we experienced what is probably the migration of the most hammerhead sharks I have ever seen in these waters. There were perhaps 3,000 of them, though it could also have been 5,000 or more. Yet, when we went down again in the same place on the reef perhaps two hours later, there wasn't a single predator to be seen.

'For the next month, sharks were scarce … hardly any … the longest period without sharks in all my time at the Banks.

'Interestingly, it was in 2008 that we had the most raggedtooth sharks here, with something like 500-plus, and they stayed with us for a long time. I actually estimated something like 300 raggies on my best dive the following year. The next best came a year later, with fractionally less in terms of sheer volume … I counted 250 on one dive in 2010, and in 2011 we once had more than 130 raggies on a single underwater excursion.'

But this still doesn't mean that the numbers are dwindling, says Mauz. 'It could have been that I wasn't around when the larger packs arrived … who can tell?'

With zambezi sharks – or bull sharks as they are known abroad – the most he has ever had on a baited dive was 15 during 2010.

Ultimately, he explains, 'we started to bait in January 2007, for no other reason than we had less sharks than before. Longer periods of zero sightings practically forced us to help nature along in order to produce the goods, something our clients were paying good money for.

'Very soon, we realised, baiting is also lucrative. Let's face it, divers don't mind paying US$150 for a dive if it means they get to take away photographs of themselves swimming in the water almost within touching distance of a massive tiger shark or a bunch of zambies.

'And it is the same each time. After the sharks casually check us out in a normal baiting situation, they tend to accept us within their sphere of underwater activity, and they do so almost as equals ... how else to explain it?

'They actually play with us, and frankly, you cannot imagine this experience if you haven't done it for yourself ... absolutely fantastic ... the ultimate privilege.'

Clearly, Roland Mauz takes his shark diving experiences very seriously and it shows. He doesn't have any kind of regard for what he terms is 'macho action'. These acts, he states, 'are simple acts for little guys,' and he goes on: 'Heroes in my eyes are the divers who manage to withstand the urge to touch a shark, to perhaps ride its dorsal fin or turn one over in the water so they have a great story for the pub back home.

'In my book, a hero has a sense of pride and respect and he or she needs to live by it.'

Through it all, diving at Protea Banks demands the implementation of certain guidelines and these have to be observed.

Fundamental boat and underwater disciplines are essential. Also, this is not a beginner's experience: in the old days you needed to have at least a master diver's or an instructor's qualification to venture down on dives which can sometimes start at 30 metres. The ocean floor slopes upwards towards shallow depths so the dive gets easier as you progress. But the currents can be powerful, awesomely so.

I'd first heard of Protea Banks from various dive enthusiasts quite a few decades ago. While I'd encountered sharks often enough in just about every ocean – including my share of great whites – I'd never 'done' Protea. That soon changed.

In the milieu of shark diving, the experience was unusual in that the place reflects none of the quiet formality of some of the more familiar shark haunts of the Caribbean or the Red Sea. Or even the nearby Aliwal Shoal where raggedtooth sharks are a feature. Regarded by many as the ultimate shark dive, Protea Banks has more sharks than almost any other comparable reef along that stretch of the Indian Ocean.

Because depth at Protea Banks can be a tough option for some divers, there was good reason for caution. None of us knew

Beulah Mauz coordinates dive instruction courses at her and Roland's operation in and around Protea Banks: African Dive Adventures is able to qualify aspirant divers within two or three days. They also offer above average accommodation at their dive lodge in Ramsgate. (Photo: Author).

anything about sharks except that they were dangerous. We were warned that on entering the water, there was to be no hanging about on the surface. Immediately on exiting, we were instructed to shoot directly down to the bottom. Only then did we check our gear and, of course, each other, something most resort divers would ordinarily have done on the surface.

Our contact with sharks that day was almost immediate. Within a minute a bunch of them loomed out of the murk and headed in our direction. Limited visibility of about three metres precluded us from seeing how many were in the pack.

Water with perhaps three metres of visibility was another of the problems we encountered that morning. Since the seas around the Protea Banks are not tropical and the Mozambique current that surges down the East African coast can sometimes touch three or four knots, we needed to concentrate both on depth and remaining in touch with one another. There was no shark cage in the event of an emergency.

As with white-water rafting and kayaking in those distant days, Protea Banks had already been classified by the majority of resort divers as challenging. British national Graham Powell, our divemaster that day, went one better: 'It can be a little dangerous at times … don't let anybody tell you otherwise,' he suggested.

From the start, our group that included five experienced scuba divers from Europe seemed to attract the attention of a family of large zambies, some aggressive and increasingly erratic as the dive progressed. 'They circled us repeatedly, steadily tightening the gap,' Graham recalled when we spoke about it afterwards, and here I'm talking about the late 1980s.

As Graham explained, some of the critters displayed the classic shark attack mode: lowered pectoral fins and rounded back. 'It's then that you take notice. But you don't change your pace, your mode or your actions,' he declared.

'Truth is, if you remain totally calm and don't demonstrate fear, most sharks will ignore you: essentially, you become like any other denizen in the sea ... they will come in, have a quick look and veer away,' he explained. In a sense, he said, it was like playing chicken. The only difference was that some of the participants had intentions, if not hostile, then a little scary to the uninitiated and that concerned him more as the dive progressed. Also, the creatures had become markedly more animated and when that happens, it was time to go home, he decided.

Once the final pair of divers had surfaced alongside the waiting boat – they kept their legs close to the hull as a precaution – Graham was left alone in the water at a depth of about three metres while still equalising pressure. Quite suddenly and unexpectedly, he became a target. Like a bunch of hounds, a dozen sharks came at him from all sides and below. In the end, he was forced to use the reel that fed his line to the float marker for protection. Several times he beat off attacking sharks using the plastic spool for protection.

'I reckon I was pretty shaken by the time I made the boat. But at least I was safe!' Like some of the others, he hauled himself on board in a single agile move: his tank and weight belt were still strapped on.

'Had I lost my nerve, I don't know what would have happened,' he told us afterwards. Also, it was not the first time it had happened, though he admits that circumstances vary. The worst-case scenario, he told me, was having someone in the dive group who displayed real fear.

'Sharks are predators. They pick up on fear, almost as if they can smell it, even underwater, which is symptomatic of all predators, on land or in the ocean.' It's instinctive, he added, because this is their milieu and they rarely miss anything.

Research has since shown that since most wild creatures – and predators in particular – do have a sense for the kind of electrical impulses that are generated by fear. It is exactly the same with distress, age or incapacitation: all part of the process of natural selection – the weak yielding to the strong, the infirm to the young. 'Terror generated in this manner is a powerful magnet for attack,' he told me.

On my first dive at Protea Banks, once on the open sea, our dive team was totally on its own. The rule from then on was standard: if one of us had a problem, everybody in the group had a problem. We were all then involved in trying to put it right and there were to be no half measures, Graham stressed in his pre-dive briefing.

'You're as one when you enter the water and you remain that way until you're on board the boat again.' More important, he said, 'nobody gets separated from the others. Sharks, sensing distress, can appear within seconds.'

The lesson was basic. It had everything to do with our survival, he urged. Also we were some distance from the coast so our only support lay with the boat which followed us from above, using the buoy which Graham towed behind him to keep pace with us. Earlier, it had taken us about 20 minutes to reach the banks on a fast, semi-rigid hulled boat fitted with two 85 hp Yamaha outboards. To cap it all, things were made a little difficult by the weather: it was overcast and windy.

Once in deep water, as previously planned – all members of our team formed a protective defensive circle before moving on. It was hardly back-to-back, but near enough to close ranks should a shark become threatening. In our case, we covered all approaches.

As Graham had instructed, when one of us spotted something coming closer, we were to make eye contact and use our arms to indicate direction. In that manner, other divers could be made aware of any kind of potential threat. In any event, he stressed in his pre-dive lecture, once spotted, most sharks back off, though it wasn't always so. Both Graham Powell and Andy Cobb have always been strong proponents of the precept that every dive in hostile waters requires planning.

On a personal level there were several set procedures. For instance, said Graham, someone who couldn't equalise on the way down, or perhaps had equipment failure, needed immediate help: he or she could not be allowed to become separated from the group. 'Therefore, when that happens,' he told us, 'everybody has to head for the surface where the problem needs to be sorted out before continuing.' If time and depth allows, the dive can again be started from scratch.

Separation from the main group does sometimes happen, as most instruction books will tell you. It is also the ultimate nightmare of any divemaster, especially when operating off the South African east coast.

The warning given is standard: lone divers face an element of peril, not only from what some people like to term 'rogue' sharks, but also from having to contend with currents and, if the seas are rough, remaining afloat. Currents can be strong enough – as has happened in the past — to whisk divers long distances down the coast before help arrives. Exactly that took place shortly before we went down on Aliwal Shoal a few years ago.

That group involved, broke the seminal rule of not staying together on the dive, or at least within signalling range. Normally, as one instructor explained, when it happens, the entire dive group will surface, report the situation to shore by radio, and they will set about initiating a search. If things are deemed really bad, the call will go out for a helicopter to join the search. At the time of writing, this service is customarily provided by the South African Air Force, though nobody is sure for how much longer, because the SAAF is not what it once was.

But this time it wasn't quite that simple. When the group eventually did get back to the boat — with a string of sharks in tow — a young Austrian couple that had originally set out with the group and had been filming, was missing. Matters weren't helped by visibility: it was down to a basic three metres. The divemaster in charge of the party admitted that he might have been more vigilant because twice in the first ten minutes of the dive, he had warned the pair not to wander off. But the recalcitrants obviously had ideas of their own.

'Once back on the boat, we got on the blower and immediately radioed the shore station. They, in turn, alerted Durban. Forty minutes later an Air Force helicopter hovered over our position.'

As he tells it, the pair was eventually found, but nowhere near where they had disappeared. They were finally lifted out of the sea about 20 kilometres from their original position; the current had shifted them another four or five kilometres offshore. As one of the more experienced divers commented afterwards, it was close. Yet the two weren't any the worse for wear. 'In those four hours,' said the man in charge, 'I aged ten years!'

Recalling the incident afterwards, this veteran diver said that once ashore, the pair was astonished at the fuss they'd created. They told their rescuers that after filming some of the zambies that were following them, they completed a long decompression stop and were accompanied by three more bull sharks all the way to the surface. Had they not taken with them a blow-up surface marker, the chopper pilot said afterwards, they would almost certainly have been lost.

Interestingly, in a previous incident at Aliwal Shoal, another diver became separated from his group. Seventeen hours later, in the middle of the night, he eventually managed to clamber ashore on a deserted beach. Like the others, he too was unscathed, even though he'd made land about 50 clicks away. Not so lucky was a group of half a dozen European divers who became detached from the main group in the Red Sea some years before and who were never seen again.

The sharks to be found along both Aliwal Shoal and Protea Banks come in all sizes and species and, as Geremy Cliff points out elsewhere, two metres are about the norm.

One of my most memorable dives involved a huge school of migrating hammerhead sharks – thousands of them – and very similar to what Roland Mauz experienced more recently.

This moving mass of hammerheads moved slowly and deliberately over our heads while Andy Cobb and I were down below at Aliwal Shoal. It was an early-morning dive and we had barely touched bottom at about 20 metres when the sea above us was clouded by this extraordinarily ordered column that could easily have been a kilometre long. Crouched on our knees, we could only stare in bewilderment at one of the mysteries of the ocean as this long line of sharks – some of them immature pups, others monstrously mature – continued silently on their way.

Most of these denizens ignored us, but some of the more immature youngsters shot down in a bid to inspect us interlopers and then, moments later, they would head back to the throng.

Who knows what would have happened if any of us needed to surface because of an equipment malfunction. Probably nothing, in retrospect: I still have to hear of a hammerhead attacking divers in those waters. Curiously, neither Andy nor I ever experienced the phenomenon again. Only afterwards did we realise that we'd been so captivated by the event that none of us had bothered to take a picture.

The boat launch area for Protea Banks is at the local NSRI station and is a comfortable drive from Durban, though with traffic the way it is these days, you should allow at least two hours and more if you're heading for the airport.

In turn, Shelly Beach's 'launch pad' is about ten minutes by road south of Port Shepstone and Protea Banks is a short hop by boat from there. It has always been a great angling spot; the reefs seem to attract great numbers of fish, which, in turn, are followed by the sharks. That completes the cycle.

Game fish likely to be encountered underwater include king-fish (ignobilis, yellowtail and big-eye), bonito, tunny, king mack-erel, kaakop (jobfish) and sangoras. While plentiful in the early

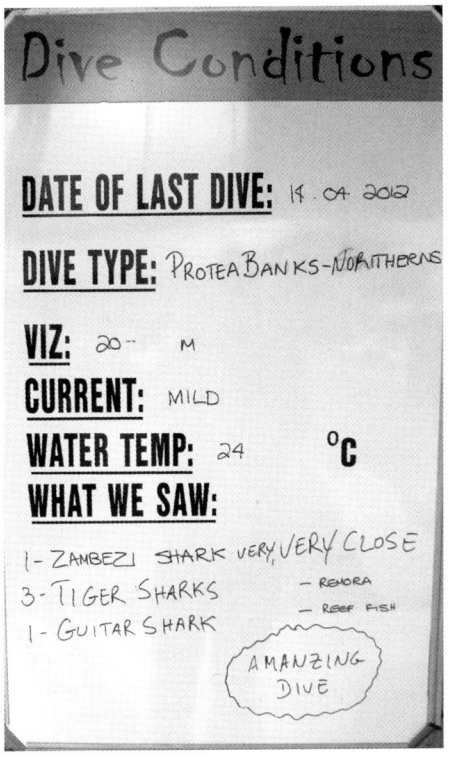

Dive Conditions

DATE OF LAST DIVE: 17.04.2012

DIVE TYPE: PROTEA BANKS-NORTHERNS

VIZ: 20- M

CURRENT: MILD

WATER TEMP: 24 °C

WHAT WE SAW:

1- ZAMBEZI SHARK VERY, VERY CLOSE
3- TIGER SHARKS — REMORA
1- GUITAR SHARK — REEF FISH

AMANZING DIVE

months of the year — before the sardine run — they taper off again and there are not as many big ones as before. One advantage is that this slots in well with Aliwal Shoal where raggies arrive in numbers about mid-year.

The most striking event involving gamefish occurred early in 1995 when Jerry Tomkins joined Roland Mauz for a dive on Protea Banks. With a lot of experience behind him, he was comfortable in the water when we sighted our first eight zambezi sharks; we had descended to the caves and were in the process of moving towards the Pinnacles.

Suddenly a shoal of well over a hundred ignobilis came into view and almost all were large and in the 15–30 kilogram range. Then they started 'dive-bombing' us, under the circumstances, a most unusual action. Talking about it afterwards, one of the wags on the boat commented that they probably linked us to the same species as the guys with rods and reels on the boats that had been trying to catch them.

The fish approached at speed and in single file, all about a metre apart. They blasted their way past us and onto the Pinnacles ahead. Visibility was about 35 metres, so it was actually quite spectacular. They regrouped and started the game all over again. An occasional shark would come into view while the fun was going on and the entire experience was videotaped.

Jerry says that it was the best dive he had ever done, which was saying a lot because he'd dived all over the world.

Roland and Beulah Mauz are old friends and operate African Dive Adventures in Margate, adjacent to the banks.

In the early 1990s, recalls Roland, 'some brave scuba divers pioneered the reef, which was then only used for fishing. But since then, things have taken off. And as shark diving became more popular, we started to push the limits.

'We have learnt so much about sharks in the interim that we are able to offer exceptional diving experiences to those enthusiasts who like to mix with these predators. That and depths in excess of 30 metres – as well as the current we experience on Protea Banks – all go together towards the definite advanced dive.'

As he states, Protea Banks can be dived all year round, with different seasons bringing different species of shark. Indeed, it is not uncommon for divers to see up to seven different species of shark on a single dive, though this usually happens during the change of seasons. His baited shark dives are now attracting a lot of attention and in this regard, he has shifted the focus in his direction from similar facilities around Umkomaas and Scottburgh further up the coast.

It is worth noting, says Roland, that on Protea Banks there had never been a shark accident – or, to be more specific, an actual attack – not in all the years of diving. 'Although the divers have no protection at all, there are strict rules of conduct that are detailed and controlled by the divemaster before and during the period underwater.'

As he maintains, the banks open up new horizons for those divers who have seen a lot, but have never experienced sharks …

The activities of African Dive Adventures can be viewed on their website at http: www.sharkdiving.co.za. You may contact Roland and Beulah Mauz at afridive@venturenet.co.za. Their domestic or international phone/fax number is +27 39 317 1483.

CHAPTER THIRTEEN

SHANE BREEDT ON SPEARFISHING AND SHARKS

Durban's Shane Breedt has been spearfishing in what others prefer to call 'shark-infested waters' for two decades. Visit him at his office and factory in Durban where he produces some of the best spearguns on any continent – it lies in the shadows of the famous rugby stadium that the Sharks call home – and chances are he'll imbue you with his love of the open water.

S harks – the predator type – usually feature in some of these discussions; his views about the creatures are hardly dismissive.

They are always around in the warm seas of the southern Indian Ocean, he'll tell you, the only difference being that he has had very few run-ins with the predators over quite a few years of spearfishing. He can count on the fingers of one hand the number of times that he has been forced to return to his boat because of aggressive shark action.

'That alone tells you something', he says. 'Each time it happened I had something in my possession that the sharks – clearly hungry – wanted: fresh fish that I'd just shot, in all probability trailing the kind of blood and muck that would attract them in the first place.

'So it is pretty much taken for granted that if you shoot fish in the open sea, your actions are not going to go unobserved by other marine creatures that live there. If it is not sharks that show interest, then it will be the occasional porpoise or even the remoras that hang around when there are sharks about.'

As for danger, he is explicit. There are several precautions that need to be taken, he advises, the first being not to dive in low-visibility water, which would include all river estuaries and the kind of dirty water found in some African harbours along the east coast. He warns that over the years, there have been several shark attacks in the approaches to some Mozambique ports as well as in the vicinity of Mombasa in Kenya and Tanzania's Dar es Salaam.

'Almost always, when this has happened, the water had low visibility,' he states, 'which tells you just about everything … and the seas off KwaZulu-Natal are no different.'

Darrell Hattingh, one of the most experienced spearos in the world, has his own take on spearfishing and sharks, which might be expected after more than 30 years of practising the sport. His views tend to slot in tidily with what Shane Breedt has to say, though clearly, he has covered a lot more water and has had his share of shark 'experiences', a word he prefers to use, rather than talking about 'attacks'.

This veteran diver is outspoken about the fact that sharks have been an integral part of the underwater environment for the spearo for as long as the sport has existed. As he suggests, once you have a shark experience during a dive, it becomes very much a part of one of the most adventurous sports there is. At the same time, he maintains, that should not detract from the fact that these predators play a vital role in the ecosystem of the oceans – which gives good reason not to harm or kill them.

He goes on: 'The bottom line as far as sharks and spearos are concerned is that if anybody wishes to become a serious spearfisherman, he or she needs to know that encounters with sharks are inevitable. If that possibility is not acceptable, then it perhaps is time to try golf or tennis instead.

'Nowadays, sharks are under enormous pressure in their own natural domain. They are being threatened by mankind like never before, to a point where they have been almost wiped out in some regions, for no better reason than commercial gain. He offers a friendly warning that the days of 'monster sharks' that freely roam the oceans are long gone. 'With the exception of sharks migrating, these overlarge creatures are limited to a few places like Dyer Island in the Southern Cape or Dangerous Reef off South Australia, both tourist destinations today and home to several large families of great whites. But even there they have become rarities – count yourself lucky if you see a great white of four metres or more.'

Human perception of sharks and their behaviour, says Darrell, has also changed. Thanks to educational programmes, rudimentary maritime awareness as well as conservation measures, the shark is better understood these days than before. It is no longer the perceived 'killing machine' it was once portrayed to be.

'Despite protestations, this dreadful image persists, which might be expected when the presence of just about any shark off a popular beach results in the kind of hysteria that makes for front-page news … even an occasional basking shark off British beaches continues to make headlines.'

Still, he reckons, with technology advancing, and spearos and divers alike exploring new and remote diving areas, sharks will always be there, at the top of the underwater food chain. 'And since we are intruding on their environment, there are always likely to be confrontations between man and shark.

'In my decades of experience as a spearo and many, many hours spent in the sea, I have only had serious encounters with a single great white, three tiger sharks, numerous zambies (bull sharks), as well as quite a few blackfins, hammerheads and raggies.

'Of these encounters, only two were life-threatening – other than being "rushed" twice by zambezi sharks.

'Perhaps the closest I have come to being chomped was at Mbibi, north of Durban, and that happened some years back. I was followed by two blackfins and a rather large zambezi shark of about 150 kilograms while returning to shore with a stringer loaded with fish that I'd shot.

'Throughout my swim back to shore, the sharks lurked at the edge of my visibility, right until the moment I was ready to swim in through the breakers. In a rather bizarre incident, the two blackfins made a couple of quick passes and their intent was obvious – they wanted my catch.

'As I turned to fend them off with my speargun, the zambezi shark homed in and went into a frenzy: it grabbed the fish on my buoy, perhaps a metre in front of me. At the same time its tail "slapped" me hard in the face and almost dislodged my mask. Undeterred, though obviously I was taken completely off-guard, I thrust my gun at it, hard! Even then it made off with a stack of my precious fish almost halfway down its throat and still attached to the buoy.

'Somehow, I managed to pull the by-now-mutilated catch from its mouth and hauled the buoy back toward me. By the time the shark came in for another bite, I'd already reloaded my speargun and managed to attach the Powerhead. The shark came in close, and with all the commotion, I got my shot off but I actually missed!

'This time its snapping jaws just missed my legs and by now I was smashing my unloaded gun into its head and, as I remembered afterwards, the fact that there was blood from the dead fish all over the place didn't exactly help either.

'I was aware that the situation had suddenly become quite dangerous. And while I'd started off in a mindset that was "fight rather than flee", I accepted that it would be stupid to go on. With that, I left the shark to finish what it had started and slowly made my way back to the shore. By the time I got to the shallows, I was left with only the head of another of the fish I'd speared that day. Score: Sharks three or four: me zero …'

As Darrell explained years later, the incident clearly demonstrated that the shark was after his catch. He might have been wiser to avoid the confrontation, but then there are dozens such stories at every dive club in the country: spearos unwilling to forego their catches because of a bunch of sharks poking about.

'The other time was when I provoked a rather majestic raggedtooth shark at Plettenberg Bay by trying to shoot it with what turned out to be a "dud" Powerhead. After the third attempt, it went for me, snapping furiously. I backed off, tempers seemed to cool and we went our separate ways …

Darrell Hattingh warns that sharks are not the only critters in the sea that bite. In a recent summer he was spearing along Nine Mile Reef at Sodwana Bay.

'After a quiet spell, a shoal of barracuda came into view and I targeted a fine specimen of about 14 kilograms. It was something I'd done often enough before. But this 'cuda was different. It promptly turned on its tail and shot in straight towards me at speed and sank two pretty formidable rows of teeth into the back of my thigh.

'Though it cost me a bunch of stitches, the incident demonstrated quite forcibly that day that sharks aren't the only critters in the ocean with bad reputations.'

Shane Breedt also has strong views on the issues faced by spearos. In answer to questions, he made some valuable pointers to aspirant spearos. All these events are based on his own experiences while diving and his views are not only instructive, but could ultimately prevent injury, or even save lives.

Asked to name the first rule that applies to each and every spearo when there are sharks present, he was unequivocal: 'You never dive alone, ever!'

The fact that there have been people who have done so in the past, he states, displays a blatant disregard for both safety and survival. It also reflects a rather short-term approach towards the sea, which can be treacherous.

'Lives have been lost because of such stupidity … you always

Morné Hardenberg

Fiona Ayerst

Sijmon de Waal

Sijmon de Waal

Sijmon de Waal

Sean Botha

Sijmon de Waal

A sequence of shark photos taken by Morné Hardenberg, Fiona Ayerst, Sijmon de Waal and Mike Rutzen (opposite page). Bottom right is Sijmon's wife Debbie in the Arabian Gulf on a night dive with a whale shark. Opposite, bottom, is Mike Rutzen with Hollywood's Matt Damon after a shark dive at Kleinbaai.

Mike Rutzen

Mike Rutzen

Mike Rutzen

A marvellous selection of Sijmon de Waal's undersea shark images in South African waters and in the Middle East. The photos above were taken off Aliwal Shoal and include a large tiger shark. Below, his wife Debbie stretches out her hand towards a great white shark while diving in Mexico.

SARDINE RUN

A selection from Mozambique with dugong pod (left) and the beautiful Nyati Beach Lodge (top right), Big Blue spearos and their catch (below); the crashed microlight partly salvaged from the sea at night by Janneman Conradie, and the man himself with Dr Andrea Walters, heading out from Tofo (bottom right).

Sijmon de Waal

need a spare pair of eyes watching out for unwanted visitors,' were his words.

He also warns that if there are sharks around, the spearo must not only be aware of their behaviour but should also be sensitive to their reactions, especially if fish have been bagged. If that behaviour changes, he suggests, move to a new spot. 'Diving with sharks is generally not a good idea.'

Though he has been spearfishing for almost all of his life, even before he left school, he has never been an enthusiastic competitor in annual spearfishing championships. 'It doesn't particularly turn me on, I don't see merit in going out diving and shooting fish to prove a point ... it simply makes no sense ...'

What he does not say is that his approach has not prevented him from going into business and producing some of the best spearguns available. Years of research, trial and error have gone into this work and his guns are today in demand all over the globe.

The best spearo venues? Shane is reluctant to be drawn, though he does admit to finding some of the best diving along the country's Wild Coast, on the northern fringes of the Eastern Cape. There are lots of good quality fish available there, almost for the taking. Also, the range of species is remarkable.

On encountering sharks in previous years, he feels that there are possibly a lot more of these critters around these days than before. 'I'm not sure why this is so, but we see more of them when we dive these days, which is contrary to what is going on elsewhere in the Indian Ocean.

And though he has been lucky with sharks, some of his friends have had some serious encounters.

'My old pal Len de Beer – he has since moved to New Zealand – was nearly eaten twice by a great white while diving at High Points off the Zululand coast. Another was James Vine with a tiger shark off Scottburgh, and then Jaco Blignaut ended up being pushed about by a zambezi shark on the surface of Deep Cracker, which also lies off Scottburgh. Not long afterwards, Alan Fraser was buzzed by zambezi sharks at Protea Banks, but he was spearfishing and not diving with tanks. The list goes on.

'I've actually been quite lucky. I have had my share of close encounters, but nothing that would keep me out of the water. We recently lost a young diver at Cape Vidal who was attacked by a zambie and lost his leg: he later died. We also had someone searching for crayfish mauled on the lower South Coast by what was presumed to be a zambezi shark, but fortunately he survived with some really nasty bite wounds.'

Asked about his most memorable shark moment, Shane Breedt recalls diving off Durban's Bluff one day in about 15 or 18 metres of water when he was rushed by a big zambezi shark.

'I was lying in a pothole stalking a fish when this shark took an immediate interest in my activities and came at me ... just like that. It was quite frightening at first, but I collected my thoughts and swam slowly to the surface. It was very aggressive to begin with and then the shark simply disappeared ... never saw the bugger again ... not really sure what it was about ...'

What he admits to doing though, was following his own best advice and getting back on board his boat and moving to another spot.

His views about the aggressive nature of larger sharks are the same as those of most other divers. He rates the great white as the deadliest and the tiger shark as the sneakiest. 'Tigers like to creep up on you, and that can be a problem if the water is dirty ... they can see you long before you are aware of their presence, which means that you really have no idea that they might be there in the water with you.'

On the question of letting the younger generation dive in the open ocean, Shane Breedt is specific.

'Letting your kids dive while under adult supervision is fine, but it should mean exactly that: *supervision*. You don't let them into the water if there are sharks about. It is as simple as that.

On equipment, this spearfisherman says that even in warmer water he uses a full suit that consists of a hooded jacket, no zips, and pants, of course. For the rest, he prefers a low volume mask, carbon fins and his one-metre Free Divers carbon gun: only the best.

Over the years this experienced diver has played a major role in the evolvement of equipment used to shoot fish. None of it 'explosive', he stresses – and not at all like normal firearms. They are called 'spearguns'. That said, he admits to these instruments having become much more advanced with time.

'The emphasis today is on how the device looks and on new additions like "rails" that are now seen on the barrels in bids to make the guns more accurate. Their triggers have also evolved and are very smooth when fired.'

Ultimately, he suggests, it is the environment that allows the average spearo to prosper, and here, he exclaims, South Africa is blessed with its almost unlimited stretches of unpolluted and largely undeveloped stretches of coastlines.

'South Africa has the potential to produce world-champion spearos

Photo: Gérard Soury, courtesy of Walter Bernardis

and we've had a few. We've actually got an array of really outstanding divers in this country at present, some of them world-beaters. On the downside, the sport is slowly becoming beset by irrational laws and regulations ... there are now too many restrictions in place and that is likely to have a decidedly negative effect on spearfishing in the future.

'This is sad, because while there are people in government who maintain that spearos are plunderers of maritime resources, spearfishing is actually the most sustainable form of fishing. We shoot what we see, so it is a very selective form of hunting.'

He points to controls. In the old days, spearos could take 32 fish per person. That has since been limited to ten. 'But there is talk at government level of limiting it further, to possibly two fish a day per diver.'

For all this, spearfishing is a growing sport. Many more young people are getting into it because it attracts youngsters with ability, strength and natural athletic ability. They want to go out there and match wits with large fish in their own milieu. Also, it tells you something that lots more young people are into it, if only to put a fine plate of fish on the family dinner table.

And it is not expensive, he maintains.

'In dollar terms,' he says, 'it would cost the aspirant spearo less than US$1,000 to set him or herself up to get out there and do their thing and that includes all that would be needed – speargun, wetsuit, fins, mask and the rest. There are many clubs where a prospective enthusiast can learn, which is the obvious route, because divers tend to listen to what others are doing, or at the very least, discover where the fish are shoaling.

'But if you are going to shoot fish, please don't do it the American way. The average Yankee underwater enthusiast will don a scuba set and go on the hunt for the big one and that is certainly *not* sport.

'As we all know, fish are inquisitive. They tend to approach scuba divers, which means that shooting a beautiful grouper or wahoo with a tank on your back is almost equivalent to taking a high-powered rifle and going into the paddock to shoot a cow ...'

Siimon de Waal

CHAPTER FOURTEEN

SHARK DIVER SIJMON DE WAAL LOOKS BACK

Sijmon de Waal, whose underwater images are prominent among the pages of this book, has arguably had more shark experiences than most. His idea of fun is the kind of short excursion he and his wife Debbie made late 2011 when diving off Mexico. Halfway through, they spotted a great white shark. No sweat, they both went into the water to record the occasion. Debbie can be seen diving with that shark in the general colour photo section.

As an engineer involved in the oil industry, Sijmon's life is hardly one great adventure. But he's had a few interesting experiences, including the one he refers to simply as Isabella, the name given to a female tiger shark.

As he tells it, he was working south of Durban early 2007 when they captured a small female tiger shark that was destined for the Ushaka Aquarium in Durban. She was not the first of this specific shark species to be captured and displayed there, the idea at the time that they would keep her in captivity for a few months and then release her back into the wild.

'Like dolphins that are held in captivity, I believe that observing such a creature in a simulated natural environment tends to act as a powerful educating medium for the public at large; most people never have the opportunity to see them in the wild.'

Isabella, he recalls, was one of the marine creatures he was to become thoroughly familiar with over the next few months. But it wasn't quite as simple as it sounds.

'It might look easy to capture a shark, but an awful lot of effort and expense sometimes goes into such an activity. Mark had developed a specific technique for capturing the animal, which involved getting her accustomed to the bait stem and then "simply" hooking her in the corner of her mouth with the help of a short pole. We'd then get her on board the boat as soon as possible and put her into a holding tank to reduce stress. Though it might be regarded by some as a lengthy process, it was actually a far better method than using conventional means to fish for the shark.

'This system required a lot of time spent in the water and for several reasons. First we had to choose a predator of the right proportions for the aquarium, and then we had to work to get her accustomed to contact with humans. Only then would we proceed with the capture. With those many hours spent at sea with the sharks, it didn't take us long to form bonds with them and to recognise that each animal – just like us humans – possesses a unique personality and that we could also have good days and bad days.

'After a successful capture, Isabella was displayed in the main tank at Ushaka for a few months and, when the time came, she was returned to the ocean. Within a week she was back on the bait stem off Aliwal Shoal south of Durban and subsequently, many months were spent with us observing her actions.

We soon discovered that her brief stint at Ushaka had actually made her less wary of humans and it was easy to interact with her. This made her an excellent ambassador for her species to the many divers that got to see her and she was no worse off given her recent aquarium experience.

Unfortunately, as with many other animals of her species, disaster was about to strike …

'We received a call that one of the well-known fishermen from Rocky Bay, a few kilometres further south from Aliwal Shoal, had bagged three tiger sharks and had them on this boat. The local grapevine suggested the number was a lot higher, as some sources suggested that Rustin Naidoo, the individual responsible for this illegal activity, had also caught some tiger sharks the previous week.

'Naidoo was eventually convicted and sentenced for the "killings": and that happened because Aliwal Shoal is supposed to be a Marine Protected Area. But what made it really distasteful was that we had taken this man to sea with us on many occasions. Like us, he had swum with these animals, he'd experienced their beauty in the wild. But in the end, it would seem that a selfish short-term allure of R12 for a kilogram of shark meat (less than US$2) proved too much for this greedy man.

'After this incident, Isabella – together with several of our other regular tiger sharks – was never seen again. We spent many months hoping that she'd evaded capture and for a while, even had optimistic thoughts that she'd one day surprise us all and appear again out of the blue.

'Sadly that never happened. Although we were never allowed to identify the shark cadavers with which Naidoo had been arrested, we had to finally accept that Isabella had probably been illegally caught and killed.

I will never forget my very first graphic images of sharks. Our local church was showing a movie – in those days it was all reel and projector stuff. The event remains a distant memory, but I would have been about five or six years old at the time.

The movie was that great classic shark film made by the Australian couple, mostly in southern African waters (though we didn't know it) called *Blue Water White Death*. Only one emotion describes the almost overwhelming feeling of seeing one of the divers emerge from the cage and swim with scores of sharks. We sat there in front of the screen absolutely transfixed, fearful that he was about to be eaten.

That part of the movie I vividly recall – the rest is a haze – but I think that it was an enormous emotional response that burned the event into my memory forever. A brilliant film four decades ago, it is still a remarkable document …

Growing up in Pietermaritzburg, we were fortunate to travel regularly to the coast for our holidays. My parents dutifully passed on the generally accepted rules for swimming in the sea, which were based on the available knowledge at the time. At beaches with shark nets, they said, we were safe; at beaches without shark nets, we were allowed into the sea only up to our knees. Any deeper, they warned, and our lives would be at risk – we would surely become shark food!

This is not something unique and peculiar only to southern Africa: most beachgoers in regions where shark attacks take place still have the same mindset.

The television series *Magnum PI* inspired my first mask purchase, but unfortunately that was short-lived as it was left in the sun too long between holiday trips and the rubber deteriorated and finally tore apart. A trip to Port St Johns in 1985, when I was in Grade 8, resulted in my dad buying me and my brother a pair of mask and snorkel sets.

Looking back today, I realise that that was the start of our absolute fascination with diving and experiencing the remarkable gifts the sea had in store for us. That holiday was spent mostly in and around

Sijmon de Waal

the many rock pools that speckled the shoreline along the fringes of what we today call the Wild Coast. Unquestionably, the experiences must have had an effect on my father, because it led to him doing a dive course about a year later.

Not long afterwards, he enrolled me in a dive course, for which I had to pay back later as a bricklayer's assistant and it wasn't long before we'd done a few dives at Sodwana and Aliwal Shoal. To our surprise, there were no sharks to be seen.

Speaking to the old dive legends of the time (the 'camel man' JR immediately springs to mind) sharks were not the attraction they are today and we would be lucky to ever see one. But pretty soon afterwards, we encountered our first shark.

The event occurred on one of those crystal-clear Natal days on Aliwal Shoal, where we'd managed to arrange a dive with the Jenson brothers on their fishing boat – that was a while before the dedicated-type scuba excursion boats arrived on the South Coast, so it was a really unique event for us all.

That dive itself was a solid experience. Being young, the imagination sometimes goes rampant and once over the side and finning along the bottom, we would find it difficult not to explore every underwater feature that came into view. Though on strict instructions to maintain the dive buddy principle of never moving away from your allocated partner, I constantly left my dad's side and rushed towards them, and with such good visibility, why not?

I found a feature that resembled a cave (it is now known as the Chunnel), zoomed straight in and there it was: my very first real live shark, only two metres away.

It was not till it turned sideways and displayed a great set of jaws and that distinctive raggie dorsal that it fully hit me … SHARK!

There was no question that it was beautiful, serene and majestic – or so it should have been. Instead, I was suddenly enveloped in fear as my stomach turned. I actually believed that I faced a horrible death. Self-preservation and the flee response kicked in, which was when I swam right into my dad, who had luckily followed me into the cave and grabbed hold of my arm. By now we had ascended a few metres above the Chunnel.

Hand signals just didn't seem to evoke a response from my father, so I did the next best thing: I took my regulator out of my mouth and shouted SHARK! Try it yourself sometime and you'll appreciate that this was no easy task and several attempts were needed.

My dad hardly reacted. Instead he just nodded and indicated with his hands that we should go back down again and take a closer look.

No way! My reaction was enough to suggest to him that I had seen enough: it was time to call for back up. We summoned the rest of the dive team and returned to the rear of the cave where we hugged the bottom. The rest of the dive was spent hiding behind the rocks staring at the lone raggedtooth shark swimming in the cave.

If you ever meet my 'old man' it's one of the first dive stories he's likely to tell you and he tends to do so rather a lot.

I regard that incident with the lone raggie as one of the touchstones of my life. Call it fascination or possibly even morbid curiosity, my interest in sharks took off.

In the years after that, we dived in more locations and encountered a variety of other predators – reef sharks, zambezis, hammerheads and the rest. I devoured every book on the subject that came to hand – that and spearfishing stories that sometimes included battles with what was then still 'the dreaded zambezi' and other so-called 'man-eaters'.

To my mind, these events were all clearly life-threatening and I was not sure whether I could ever muster up enough courage to face some of the more dangerous species of shark by learning to shoot fish. There were enough stories circulating at the time (and persist today) that any fish shot would cause sharks to start circling you in the water in an instant.

Not sure why I even did it, but I and a good friend from junior school, Rob Schapers bought ourselves a pair of spearguns and instantly became spearos. Our first attempts were excursions out to Quarter Mile Reef at Sodwana Bay, where we never managed to shoot anything. The prospect of seeing a shark was always in the forefront of our minds and we were not sure how we would react and what we would do should it happen. Well, in the end we never did see any.

We had tried to get out to sea with one of the local spearo/skippers Damian Whyte. Looking back, the tiny 50 centimetre spearguns we'd bought (with no line or float) was probably not the best tool for the job when attempting to bag game fish.

Damian summed us up pretty quickly and could see here were a couple of plonkers that were sure to ruin his afternoon's spearfishing. Strangely enough, he was later to become (and still is) one of my best friends and we got to spend several years working and spearfishing together.

News of the opportunity to dive with zambezi and other sharks at Protea Banks opened up a whole new dimension as to what animals you could see if you were in the right place.

Trevor Krull started a dive concession at Protea Banks in the early 1990s and it was an opportunity we simply could not ignore. On that first trip out, Rob and I both admitted afterwards that we wondered what we were doing there. After all, everyone at the time knew that zambezis and the other shark species that we had a chance of encountering at Protea were dangerous and, surely, something was bound to happen. Initially, two spearguns were always taken along as protection.

Fishing magazines at the time warned of the impending disaster with divers now diving on the 'shark-infested' Protea Banks. It didn't take long to realise that weapons were not needed as protection and the whole fear element of the activity was replaced by amazement. The sights there in the early 1990s were really quite spectacular.

During the summer months, packs of zambezi sharks, hundreds of hammerheads, sand sharks, huge schools of tuna, kingfish and other game fish were around. During the winter months the zambezis would disappear, but were replaced by hundreds of raggies and copper sharks. Few dared to spearfish there given the numbers of sharks. Then the 'war' with the fisher folk at Protea Banks started.

Commercial fishermen, like everywhere else in the world, quickly reduced the fish population and killed off many sharks. Spearos also contributed to this decline with many sharks having been killed with Powerheads, ostensibly, to protect the spearo's *precious* catch. If one looks at it, we as South Africans have behaved no differently to the Chinese fishing vessels that continue to plunder the sea. Fish are easily converted to money, and the immediate consequence is that the only interest those who sell fish commercially have, is in dollars/euros/yen and all the other currencies-per-kilogram. Sharks directly impact on this price-per-kilogram structure, and that means that they are soon eliminated from the economic equation.

Each group blamed the other, with nobody accepting responsibility for the astonishing declines in sharks and general ocean life. Throw in the indiscriminate destruction of the shark nets and, more recently, drumlines, one wonders what will be left in the years to come? Since then, things at Protea Banks have improved and divers and the fishing community have reached something of an accommodation; at least the shark numbers have increased.

After a stint in the Red Sea (early 1996) it was really the opportunity Mark Addison gave me to work in Mozambique that allowed frequent interaction with sharks and a period where we were privileged to experience things in the ocean that few people have.

It was this constant exposure to sharks that really changed my approach attitude. Having survived many 'close calls' while spearfishing with Barry Skinstad and Damian Whyte, it didn't take long to accept that the sharks were actually just curious, and existed at the place where we were doing our diving. While we were in the water with them, these predators mostly swam around and waited for an easy meal.

Sijmon de Waal

After years of diving with sharks, we got used to diving down and finding ourselves surrounded by perhaps a dozen fairly hefty zambezi sharks. Then we'd go about our business, shoot a fish and see if we could land it, or if those buggers were going to get it.

I remember one afternoon on the Pinnacles with Barry Skinstad. You could never tell if the sharks were going to take the fish, so the policy that evolved was to always shoot first and see what happened next. Some days the critters honed in immediately: other days they got excited but did not try to take our catch from us.

The rules were simple. If you shot a fish, the norm was that your buddy would prod away any interested or hungry sharks and prevent them from taking a jawful. The reason was obvious: After one of these animals had taken a chunk of your prize, chaos would replace order and the sharks would go into a feeding frenzy. There was no controlling them in that mode and any fish at hand would be quickly targeted and devoured.

Barry once prodded a large blacktip shark, and it took immediate offence and turned on him. It promptly swallowed his speargun, doing so from its front-end. The entire thing seemed to disappear down its throat, right up to the handle, which would make for a good metre-plus. The shark was still snapping, and Barry – now without his trusty speargun – had to fend off the predator with his hands. It

Sijmon de Waal

quickly swam away with gun and buoy in tow, regurgitating it after about a minute, which allowed us to recover the gear.

The experience was both unique and unexpected. In fact, it was the first time we'd seen such behaviour.

The next day Barry was at it again and shot another fish. I dived to protect his catch. With the previous day's events fresh in my mind, my efforts were rather half-hearted when a large blacktip emerged from nowhere and bit deep into the fish's spine. Undeterred, Barry managed to pull it free and quickly pulled his prize towards him, until it was about a metre below him.

I swam to the surface and looked down. Suddenly there were a couple of dozen sharks racing in our direction off the bottom. I wasted little time in sliding in alongside Barry and, for what seemed like an eternity – but was probably only ten or fifteen seconds – we were at the heart of a pack of zambezi and blacktip sharks, each one of them intent on trying to claim a share of the bounty.

Under normal circumstances, sharks tend to stick to the bottom, but it was the first time we had seen such a crazy rush towards the surface.

It was a thumping moment at the time, but it did make us realise that if they wanted to 'attack us' there would have been very little that we could have done to protect ourselves. In reality, I have been able to conclude after many years of diving with sharks that the practice of killing these creatures with Bangsticks or Powerheads is not for self-protection, but rather stems from fear. There is sometimes also an unbridled macho element involved, with the shark killer portrayed as a hero. If anything untoward is likely to take place, the average spearo simply has no time to fit a Powerhead to the end of his or her spear.

Essentially then, the very act of bringing an explosive device – usually incorporating a twelve-gauge shotgun shell – into play, reflects a premeditated decision to use it.

Justification afterwards is always the easy part, as the audience has been mostly programmed by movies like *Jaws* (and *Jaws 2* and *Jaws 3*) or TV documentaries that are specifically geared to have the viewing public fear animals.

As we gained more experience from interacting with sharks while spearfishing along the South African coast – and later, through actively baiting for the animals – it became evident that much of what had been written about these predators was nonsense. Sharks are dangerous, yes, but not aggressively so, though there is always the exception to the rule, usually with great whites.

The fact that shark attacks happen – though not nearly as often as attacks on humans by crocodiles or hippos – and we all had our share of what the media like to call 'close shaves', nobody should be allowed to kill an animal because it represents a threat. If that were to happen globally, there would be no lions, no tigers, no leopards and certainly no bears alive today.

The same principle must apply to sharks, though the Chinese are not prepared to listen to any of this. At the same time, it is essential to comprehend the very real fear that some people experience when they encounter sharks while diving in the open sea for the first time. It's only through being actually exposed to them and experiencing 'survival' encounters that norms and perceptions will change.

The fairly recent activity of baiting sharks with chum has afforded many more people the opportunity to interact with these beautiful creatures. Without baiting, there would not be as many people interested in saving sharks as before. Indeed, it is actually quite difficult to find and interact with sharks in the open sea, and, until recently, most divers who spent time in the water, sometimes for years, never saw a live shark 'in the raw'.

During the course of our early years, we didn't have the kind of money that many youngsters have today. Consequently, an underwater camera was something we could only dream about. If truth be told, our bar bills always seemed to swallow up a large proportion of our earnings each month, and it was not until we had completed our studies and graduated to more rewarding work that we could afford the ability to capture some of the amazing underwater sights that are so commonplace today.

My first attempts to capture marine images while diving was sponsored by a friend. Only then was I able to buy a decent underwater camera. This phase lasted a few months until I was taught an invaluable lesson: that it is really not a good idea to swim about with a fish head that you intend to use for bait, together with an expensive camera and not look round to see if you are being followed …

One of our favourite tiger sharks in the 'chumming brigade' of Aliwal Shoal, one bright and sunny morning, sneaked below me and took both the bait and camera in her jaws and tried to go her way. I was fortunate not to have had my hand in the way, but after a vigorous bit of tugging, the shark ripped my housing apart and swallowed the camera whole … whoops! Sorry, Mark. Fortunately, he was quite understanding.

Sijmon de Waal and his lovely wife Debbie, both now resident in Abu Dhabi. Sijmon, one of the best underwater photographers around, turned 40 while this book was in production.

Looking back at the marine encounters we had over the years, my diving experiences have been quite remarkable. With it, I have forged friendships that will last for life.

What started out as fear and a healthy dollop of trepidation – coupled to morbid fascination for the dangerous unknown – has been replaced by awe and respect. To my mind, it is a privilege to be able to dive among these magnificent creatures that managed pretty well for millions of years until us humans came along and started to cull their numbers.

Indeed, it is impossible to describe the sensation of being alone in the water, silently surrounded by schools of hammerheads, or perhaps watch a pack of sharks head past without having to fear the consequences.

This change in attitude has only been made possible through direct interaction with the animals. Nowadays, together with my wife Debbie (it was sharks that sparked our mutual interest), we have been fortunate to travel to other parts of the globe to work with those enthusiasts who share our sentiments.

Hardly experts, we have observed first-hand the grace and beauty of these animals. We are now also aware of the way they are being killed in great numbers and that new laws should be put in place before it is too late, both for the sharks and for the ocean heritage that our children are likely to inherit. There is no question: sharks need to be protected, and soon.

The truth is that sharks and other marine animals have been placed under severe pressure, to the extent that they are being fished to extinction.

I can only hope that through the efforts of re-educating those around us about these dangers, with the possibility of getting more people actively involved with shark conservation, that it is not too late for us humans.

With an ever-expanding world population, more efficient fishing methods and the overwhelming burden of personal greed and corruption, one cannot help but feel that the race has already been lost. If everyone were to make some contribution — in whatever manner they choose — it can, and will, ultimately be possible to reverse the process.

Only time — and the next generation – will tell whether we have been successful.

CHAPTER FIFTEEN

KWAZULU-NATAL'S SHARKS BOARD

A fair proportion of the coast in the old South African province of Natal – particularly around many of the tourist resorts – is protected by shark safety gear. The only institute of its kind in the world, the Sharks Board maintains shark safety gear at 38 localities that offer safe bathing to tourists, while also conducting research into shark life history as well as a public education programme. Durban columnist Graham Linscott tells us about it.

S ince 1962 the KwaZulu-Natal Sharks Board has provided bathers effective protection from shark attack at the subtropical beaches that have made the coastline of this South African province a major tourism destination.

The systems employed have attracted both controversy and criticism, especially since this organisation uses a combination of offshore nets and drumlines – baited hooks – to reduce the shark population and reduce the likelihood of sharks encountering bathers. It is notable that the nets/drumline combination is also used in Australia.

Researchers and scientists attached to the Sharks Board are adamant that the nets do not provide a barrier against sharks entering any bathing area. They can swim over, under or around them but, they insist, all other issues aside, statistics do show them to have been highly effective since shark attacks have been reduced to about one a year.

They pointed out that so-called 'complete' barriers have been tried, but they have been costly and ineffective because the high-energy surf soon destroys them. The Board has also been active in developing alternatives to netting and drumlines, such as electric barriers.

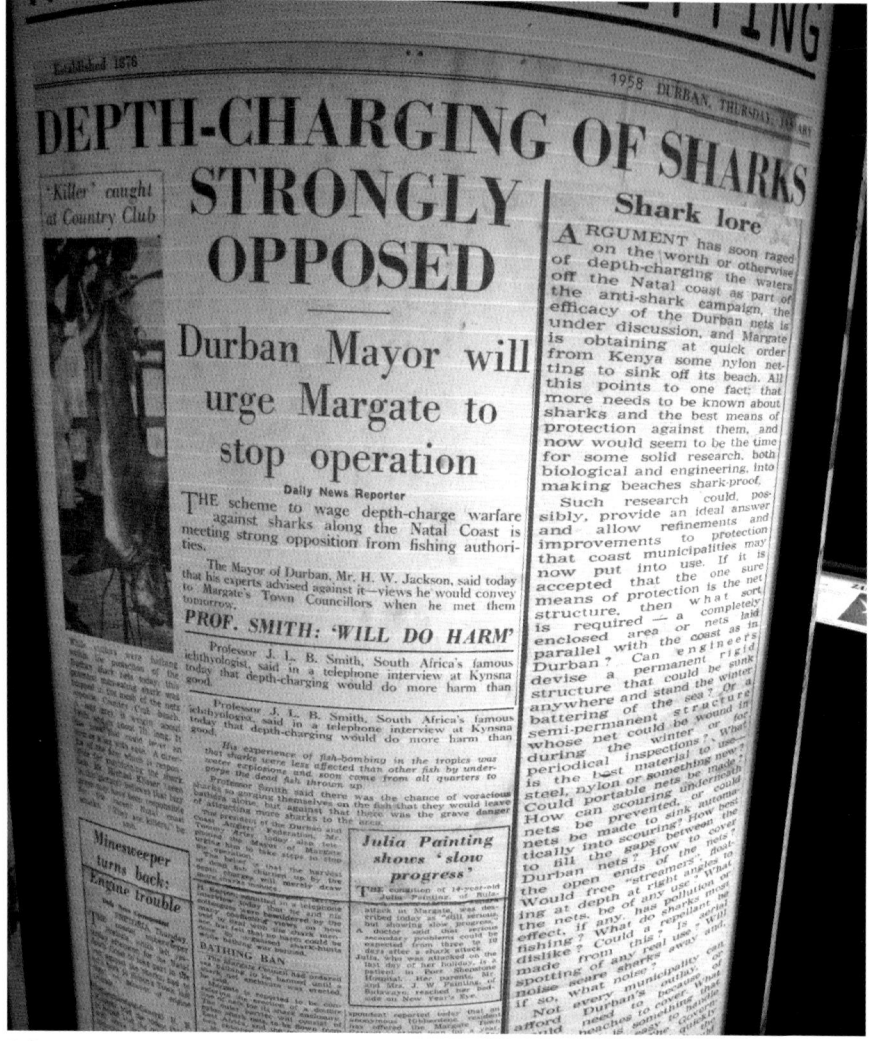

Following a spate of shark attacks along South Coast beaches in the late 1950s, much controversy surrounded the depth charging of the area by the South African Navy frigate SAS *Vrystaat,* on which the author then served on the lower decks. This newspaper headline was framed and mounted and stands in the auditorium of the Sharks Board. (Photo: Author)

Measures against shark attack began in 1907 when the Durban City Council erected a large semicircular enclosure, approximately 180 metres in diameter to protect swimmers from the surf and strong currents and against shark attack. It was constructed of steel piles with vertical steel grids between them. But the enclosure was demolished in 1928 as a result of surf damage which can sometimes be severe. Extensive corrosion and the high cost of maintenance were also factors that played a role in discontinuing that work.

For the next eleven years, shark attack seemed to have receded as a problem in Natal (possibly connected to a commercial shark-fishing operation based in Durban that was eventually closed). But in 1940, five attacks took place along an eight-kilometre stretch of coastline south of Durban; between Amanzimtoti and Winklespruit. Then Durban beaches became the focal point for shark attack. Between 1943 and 1951, Durban had 21 attacks, seven of which were fatal.

Desperate, the city authorities adopted a system successfully used in Australia since 1937. At several Sydney beaches, large-mesh gill nets anchored seaward off the breaker zone not only trapped large sharks but reduced the incidence of attack. In 1952, seven gill nets, each 130 metres long, were laid along the Durban beachfront. In the first year of operation 552 sharks were caught and no serious shark-inflicted injuries have occurred since then at any of Durban's beaches.

But these safety precautions did not extend elsewhere on that stretch of Indian Ocean coast, particularly resorts south of Durban. A series of attacks between December 1957 (for good reason, it was called 'Black December') and Easter 1958 claimed the lives of five people in 107 days. These onslaughts had a devastating effect on the coastal tourist industry and holidaymakers cancelled in their thousands.

Several seaside towns tried to emulate the old Durban bathing enclosure. Physical barriers – poles, wire and netting – were put into the surf zone to enclose bathing areas. But, once more, these unsightly structures could not stand up to the heavy oceanic pounding and were soon abandoned. The truly desperate measure of dedicated depth-charging by the South African Navy frigate SAS *Vrystaat* killed eight sharks, but the consensus today is that it probably attracted more to the area to feed on dead fish. Interestingly, Al Venter was serving as a rating on board this converted former Royal Navy warship at the time.

The logical solution was an expansion of Durban's netting operations, and in 1962 shark nets were installed at some of the larger holiday resorts to the north and south of Durban. At that time the Natal Provincial Administration created a statutory body, known as the Natal Anti-Shark Measures Board (now called the KwaZulu-Natal Sharks Board), which was charged with approving, controlling and initiating measures for safeguarding bathers against shark attack.

By March 1966 there were 15 beaches with protective nets, maintained either by commercial fishermen or municipal employees.

At this stage the KZN Sharks Board field staff worked in a supervisory capacity, but in 1974 the organisation began to take over the servicing and maintenance of net installations and eight years later it was solely responsible for shark netting in the province.

Most sharks pose little or no danger to humans who enter the sea. The number of people who drown at sea along the South African coast each year far outnumbers fatal shark attacks.

Most of the shark nets deployed by the KZN Sharks Board are 214 metres long and 6 metres deep. They are secured at each end by two 35 kilogram anchors: all have a 'stretched mesh' of 51 centimetres. It is worth mentioning that the nets are laid in two parallel rows, approximately 400 metres offshore and in water depths of 10 to 14 metres. A drumline, in contrast, consists of a large, anchored float (which was originally a drum) from which a single baited hook is suspended. Most beaches are protected either by two nets or by one net and four drumlines, but the quantity of gear varies from beach to beach.

Durban, the largest coastal city and holiday resort in South Africa, has 17 nets, each 305 metres in length, which cover all the popular swimming beaches between the mouth of the Umgeni River and the harbour entrance. KZN Sharks Board boat crews service the nets every day, weather permitting.

As we have seen, the KwaZulu-Natal coastline is often subject to large swells and strong currents, which may result either in the nets becoming entangled or the nets or drumlines being dragged out of position. Boat crews make every effort to rectify the problem when they service the equipment in the early morning, but adverse weather sometimes makes this impossible. Tangled or displaced equipment may no longer be able to 'fish' effectively for sharks and, in the interests of public safety, all bathing and surfing is temporarily banned at that beach. As soon as the equipment is restored to normal working order, the ban is lifted.

Bathing may also be banned even though the equipment is in order. This may take place if a large shark is sighted in the immediate vicinity of the bathing area or following the stranding of a whale or whale shark in the area, an event that often attracts sharks inshore, as has happened in April 2012 when a small whale became entangled in the nets off Scottburgh on the South Coast. Numerous sharks moved in to feed on the dead whale, causing 14 tiger sharks and numerous other maritime creatures to become entangled in the nets and die. This loss caused a furore, as well as strong demands for all shark nets to be removed along the entire length of the coast and destroyed.

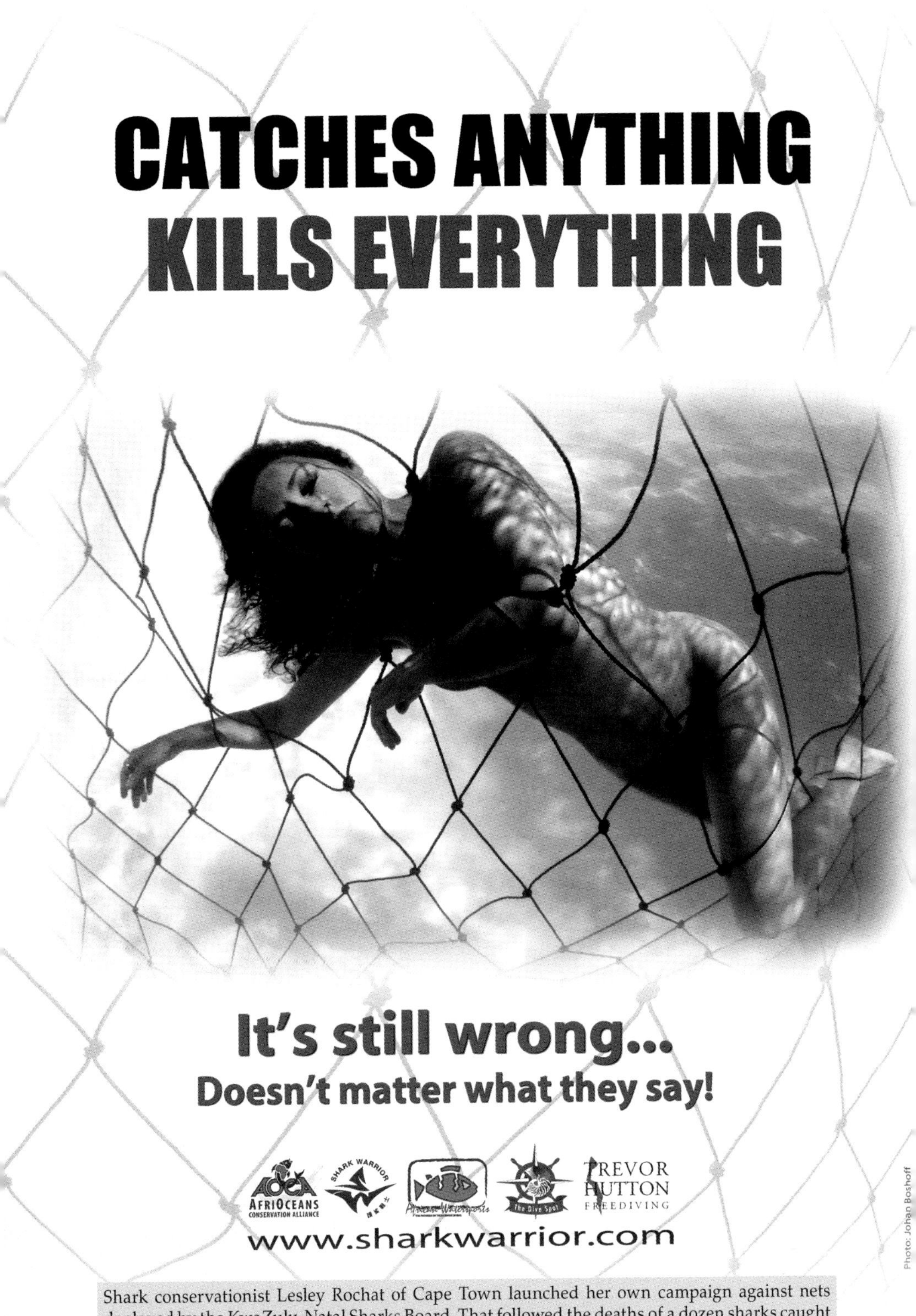

CATCHES ANYTHING
KILLS EVERYTHING

It's still wrong...
Doesn't matter what they say!

www.sharkwarrior.com

Photo: Johan Boshoff

Shark conservationist Lesley Rochat of Cape Town launched her own campaign against nets deployed by the KwaZulu-Natal Sharks Board. That followed the deaths of a dozen sharks caught up in the nets after a small whale had become entangled off Scottburgh in April, 2012.

A model of one of the great white sharks caught in the nets off Durban: it stands mounted at the entrance to the Sharks Board headquarters at Umhlanga, KwaZulu-Natal. Caroline Castell, herself a shark diver, provides a measure of perspective by doing what she'd never even contemplate were the shark still alive. The Sharks Board is well worth a visit: you can observe the dissection of sharks and possibly see Geremy Cliff at work. (Photo: Author)

An analysis of South African shark attack records over the past four decades has shown some interesting patterns. Most importantly, the results confirm that attacks on humans are rare, with an average of only six incidents per year. Since 1990 only 26 per cent of attacks have resulted in serious injury and only 15 per cent were fatal. This equates to an average of one serious shark-inflicted injury every year and one shark-inflicted fatality every 1.2 years along some 2,000 kilometres of coastline that stretches from the Mozambique border to Table Bay (Cape Town).

Initially, most attacks took place on swimmers in warm, shallow waters on KwaZulu-Natal beaches, but the shark nets – now concurrently with the drumlines – have greatly reduced these incidents in the province to less than a single attack a year.

It is significant that there have been only two serious attacks at protected beaches in the last 30 years. Both involved surfers who were bitten in very clear water by great white sharks (*Carcharodon carcharias*). The last attack at a protected beach took place in 1999, which bears testimony to the success of the shark nets in these waters.

In KwaZulu-Natal more attacks have taken place in the late afternoon than at any other time of the day, despite most people swimming in the heat of the late morning and early afternoon. This statistic is presumably related to the tendency of many sharks to venture inshore in the late afternoon to feed.

Shark attacks have often taken place in very murky waters when the rivers come down in flood. Zambezi (bull) sharks (*Carcharhinus leucas*) are known to regard these conditions as favourable for feeding and have sometimes been responsible for many attacks in water no more than waist deep.

While the number of incidents in KwaZulu-Natal has been reduced by the use of shark fishing devices, attacks in the Eastern and Western Cape have increased. Most of the recent victims in Cape waters have been surfers and bodyboarders, a pattern that has emerged in other temperate regions of the world. This is not surprising as wetsuits enable the wearers to spend long periods in the sea and wave-riders venture further offshore than swimmers.

Divers spearing fish have been another high-risk group. They operate in depths of 20 metres or more and may venture several kilometres from the shore. There is no question that blood and irregular, sometimes violent vibrations of their struggling, wounded fish attract sharks.

There are several cases of a spearo being bitten while finning to the surface immediately after spearing a fish. They often see sharks, indicating that the latter – particularly larger individuals – do not fear humans and may be sufficiently curious to venture within close range. On the other hand the incidence of attacks on these sportsmen and women, despite the temptation of struggling and bleeding fish, is less than one per year, thereby confirming the reality that in the main, sharks do not regard humans as part of their diet.

As far as can be ascertained, there has been only a single attack of any significance on a scuba diver in South Africa, in which the victim was fatally injured while swimming at the surface before descent. There have been a few other incidents where scuba divers have disappeared while underwater and were thought to have been attacked, but that premise has not been incontrovertibly proven. One mitigating factor why there have been few attacks on scuba divers is because they usually dive in groups and the noise from the continual stream of exhaled bubbles tends to deter all but the most aroused shark. In South Africa it is illegal to shoot fish while using scuba or any other artificial form of air supply.

The worst year on record for shark attack in South Africa was 1998, when 15 incidents were recorded. All took place in Cape waters; one was fatal and a further five resulted in serious injury. As the previous highest total was nine in 1994, there was much speculation in the media as to the reason for this marked increase.

A popular scapegoat was the burgeoning great white shark viewing industry, as *Carcharodon carcharias* was responsible for all six serious attacks in 1998. It was suggested that, through the sharks' repeated exposure to bait and chum that were then used as attractants – and still are – they had learned to associate boats and scuba divers with a possible meal.

The regulations imposed on the operators in this industry stipulate that the baits should be kept away from the sharks. That obviously results in limited feeding. Baiting is not done by the scuba divers from within cages, so the sharks are unlikely to develop an association between the divers and food.

Furthermore, white sharks attracted to the baits are highly mobile and may only be exposed to the chumming for a very short time. Another compelling piece of evidence against the argument linking shark attack with cage diving has been that none of the attacks took place anywhere near the shark-viewing operations. To date there is still no explanation for the spate of incidents in 1998. Indeed, apart from some recent incidents involving sharks in False Bay, attacks in the vicinity of white shark cage diving operations are rare.

Another worrying trend has been the relatively high number of incidents in False Bay and adjacent waters around the city of Cape Town. There were very few attacks in these waters prior to the new millennium.

In the 1990s there were six incidents, with a single fatality (a spearfisherman) and two other serious injuries (surfer and spearo). In the current decade, there have been more than a dozen incidents, one of the most recent involving a Springbok body surfer, who was attacked and killed by a great white shark at Kogel Bay, south of Gordon's Bay.

The most disconcerting aspect of these attacks is the fact that there were quite a few fatalities, while several other victims incurred serious injuries. Cage diving with white sharks once again returned to the spotlight as this activity is undertaken at Seal Island in False Bay and, despite the absence of any link, many suggest that cage-diving operations are responsible for the increase in attacks.

The City of Cape Town held a specialist workshop in 2006 in conjunction with the World Wildlife Fund and Department of

Environmental Affairs and Tourism to discuss the possible reasons for the increase and what preventative measures should be introduced.

Following a single, minor incident in 2007, there were no shark-related incidents in Cape Town waters in 2008 or 2009, though there have been some subsequent attacks which might have been avoided if those attacked had heeded warnings about the presence of large great white sharks, mostly in the waters that surround Fish Hoek. In part, this can be attributed to the regional spotter programme, in which lookouts situated at high vantage points alert beach users whenever they see a white shark approaching the beach. For more information visit the website www.sharkspotters.org.za.

It says a lot that many of the injuries inflicted on humans by sharks have been minor. Consequently, it is incorrect to conclude that every incident is the result of a shark trying to eat its victim. Some sharks, such as the great white, tiger (*Galeocerdo cuvier*) and zambezi sharks are aggressive species. They are equipped with razor-sharp teeth and wide mouths and are capable of causing considerable damage with little effort, but that is generally not the case. Also, there is little or no evidence to indicate that these species defend territories, but some attacks may be the result of a predator either investigating or repelling an intruder in its immediate vicinity.

Nearly 25 per cent of attacks in Cape waters (well to the east of Cape Town) have been by raggedtooth sharks (*Carcharias taurus*), also known as sand tiger or grey nurse sharks, but this creature is not regarded as an aggressive species and the injuries have all been superficial.

The raggedtooth shark has long, sharply pointed teeth that lack serrations. Such dentition is designed to seize relatively small fish that are often swallowed whole rather than to tackle prey the size of a human. Most of the incidents involving this species have taken place in shallow, murky waters and are probably a result of shark and man literally 'bumping' into one another.

There has been a decrease in fatal injuries through the development of a standard treatment procedure and the widespread availability of a specially designed first-aid kit, known as the Shark Attack Pack. This trauma pack enables treatment to be initiated on site, and is aimed at restricting blood loss and providing intravenous fluid replacement.

Essentially, the key to treatment lies in controlling bleeding and replacing intravenous fluids as quickly as possible.

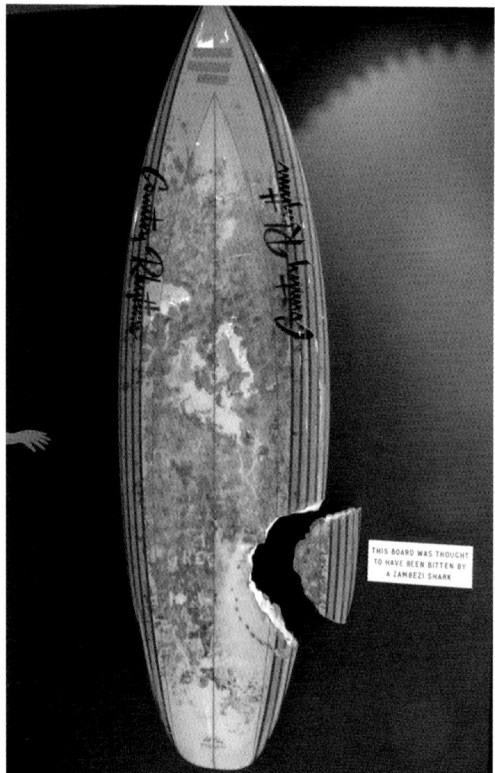

Actual surfboard attacked by what was believed to have been a bull or zambezi shark. (Photo: Author)

CHAPTER SIXTEEN

MOSSEL BAY AND ITS OCEANS RESEARCH FACILITY

In 2001, marine biologist Ryan Johnson and scientists from Marine and Coastal Management started a project using newly developed technologies called 'Acoustic Listening Stations'. These stations worked in conjunction with transmitters attached to studied subjects. The theory behind this technology was that permanently positioned 'underwater listening stations' would detect and archive the presence of great white sharks as they passed through the detection field of the listening station.

We lowered an array of eight listening stations onto suspected white shark hotspots throughout Mossel Bay and began a process of placing transmitters on great whites we located in the area. Over the next five years we tagged 94 different great white sharks in Mossel Bay, ranging in size from 1.5 metres to fully five metres!

The results were staggering. Many of the sharks resided within Mossel Bay for over six months a year, particularly during winter and spring seasons from June to November. Rather than the transient nomadic wanderers that some believed these predators were, many of Mossel Bay's great white sharks were shown to be the epitome of 'home bodies'.

The project was not without its share of hiccups. When firmly emplaced, the receivers were elevated slightly off the seafloor by using 1.5-metre iron bars that extended from concrete-filled tractor tyres. Curiously, we often relocated these corner iron bars and found that they had been bent completely flat, a mystery that puzzled everybody. The conundrum remained until we became aware that the 'flattening' process coincided with the seasonal presence of southern right whales in Mossel Bay. Our poles had become a convenient scratching post for these leviathans.

The other worry was finding divers who were willing to take the giant leap into the middle of great white shark hotspots in order to change the receivers. Fortunately, a bunch of commercial divers from the Dutch salvage company Smit Marine put up their hands. When they were not available, the duty was duly passed onto 'yours truly'.

It was a ground-breaking project that finished in December 2006 and ended with us being able to rewrite our understanding of how white sharks used this little bay, almost at the southern tip of Africa where the Portuguese mariners had first stepped ashore more than five centuries ago in their efforts to discover a trade route to the East.

The other good news is that Oceans Research has recently attracted the attention of an internal collaborative project named the Oceanic Tracking Network or OTN. This venture is currently placing massive lines of new receivers around the world to study the migration of fish around the planet. As of January 2012, we restarted this project with a number of new receivers that are not only more robust than their predecessors, but also more durable.

The future holds much promise for Oceans Research, and hopefully our whale friends will go kindly on their new scratching posts.

It was in 2001 that Ryan Johnson began conducting his doctoral research on the behaviour and ecology of white sharks in Mossel Bay, South Africa.

His initial research focused on the population structure and abundance of white sharks as well as residency patterns in the bay, coastal swimming behaviour, predator/prey interaction with Cape fur seals, together with the impact of cage diving on white sharks.

For five years, Johnson worked on these projects. After he left Mossel Bay to write his doctorate in 2006, he began to think about his long-term future with great white sharks.

Meanwhile, there had been numerous marine biologists who had approached him to ask for help in conducting their own studies, but after searching and surveying different locations in southern African waters, he realised that in terms of research opportunities on great white sharks, nothing compared to Mossel Bay.

Mossel Bay, or more strictly in Afrikaans *Mosselbaai* – mussel bay – is a protected bay known for its fine conditions, coupled to weather conditions that are conducive to conducting research at sea for at least 300 days a year. In terms of accessibility to marine animals, it is unprecedented: white sharks are found at several congregation sites about five kilometres from a protected port. Also, the amount of human influence is less than in places such as Gansbaai or False Bay.

Johnson consequently approached three colleagues, Enrico Gennari, Toby Keswick, and Steffen Swanson to discuss whether they would be interested in taking up residency and setting up a full-time institute in Mossel Bay. The intent was not only to carry on with the earlier research that Johnson had initiated, but to use the consequential baseline data to ask some really ambitious questions about great white sharks and other marine species in that area.

The new-found institute would give these four marine biologists an opportunity to set up a facility of their own, and which might incorporate their own visions and dreams. All finally agreed and after spending months establishing a structure and logistics, the die was cast: Oceans Research came to fruition in 2008.

Essentially, the role of Oceans Research is to provide and facilitate innovative and dynamic marine research relevant to the management and conservation of Southern Africa's marine life. Its research is interdisciplinary and ranges from biological to the socio-economic

Marine scientist Ryan Johnson prepares for a dive with great white sharks in the waters off Mossel Bay. As he told the author, these little excursions can be tricky if there are whites about.
(Photo: Oceans Research)

study of marine resource utilisation in Africa's coastal society. Obviously it also has a role in the country's fishing industry as, with time and effort expended, much valuable data has emerged.

The research body, as it is structured, specialises in investigating the biology of marine megafauna, including sharks, rays, marine mammals and turtles, and advises governmental and non-governmental bodies and the general industry on relevant conservation issues.

Currently, Toby Keswick, Enrico Gennari, and Ryan Johnson run the facility in Mossel Bay. Steffen Swanson has since left Mossel Bay to establish a marine ecotourism business in Cape Town.

Other key role players associated with Oceans Research include important collaborations with several South African universities. Oceans Research offers practical and theoretical training for aspiring marine scientists, from internship to postgraduate levels in conjunction with several technical colleges and universities.

Partnerships and collaborations include the University of Cape Town, University of Pretoria, the Nelson Mandela University as well as Rhodes University's Icthyological Institute. Postgraduate students enrolled in these universities are currently conducting research at the facility in Mossel Bay. In addition to in-house research projects, the institute provides facilities for visiting marine scientists to conduct a variety of biological and oceanographic projects through-out the major marine biomes of Southern Africa.

As a separate entity involving research, the great white shark is accepted as a marine apex predator that is able to maintain biodiversity through direct and indirect predation effects. It consequently represents a keystone species crucial to the performance of coastal aquatic ecosystems.

Globally, as the population metrics of the white sharks at congregation sites off the coast of central California in the United States and Mexico's Guadalupe Island have been established, pressure is now in place for South Africa to produce a variety of baseline data of its own.

South Africa is today internationally recognised as one of the world's centres of abundance for white sharks and hosts a genetically different population from those found in the North Atlantic and Australasia.

White sharks migrate along the entire South African coast and form a metapopulation between identified congregation sites such as False Bay, Gansbaai and Mossel Bay.

Incidental by-catch by fisheries, shark nets, sports fishing targeting, as well as the black-market trade of jaws, teeth, meat and fins are among the direct threats faced by the white shark. Whereas indirect threats encompass the degradation of inshore habitats as a result of coastal development, the decline in prey species is largely due to overfishing and significant public pressure to remove protection status for great whites because of human–shark accidents.

Human misconceptions regarding the increasing white shark abundance since their protection in South Africa in 1991 can be attributed to several factors.

White sharks are commonly found in near-shore locations and are known to frequent river mouths, where there is much recreational activity along the shores. It is axiomatic that as the numbers of water users increase, so too do the number of human–shark interactions. Poor journalism, where the facts are subject to question, has caused serious damage, in particular to the actions of a number of people who have worked towards the conservation of the species. There is also the issue of misinformation, usually presented to the public in sensationalist articles on shark attack incidents. Commonly the great white shark is assumed to be the culprit, when subsequent inquiry has shown this not to be true.

Consequently, these population estimates are crucial as the public's fear of an increase in white shark population is detrimental to the kind of conservation support given by the public.

The truth is that the photographic identification of white shark individuals is a non-invasive, and non-destructive technique for accumulating long-term data that covers the relevant and absolute trends in population composition. In this, the white's dorsal fin is unique and can be categorised, almost like a human fingerprint.

Therefore, identifying unique individuals necessitates looking at a variety of multiple traits. These include looking at differences in notch formation on the dorsal trailing edge, the presence of white and black pigmentation on that dorsal fin and other distinct markings or deformities as well as estimated size and sex.

The white shark was one of the first elasmobranch species on which such a technique was applied, specifically because these creatures return to specific areas at regular intervals: this makes for possible photo identification. By applying mark and recapture models to data from uniquely identified individuals, researchers are able to attain robust estimates of a population's size. Estimates on abundance – specifically ones obtained over a long period of time – are required to determine population trends, gain insight into the health of the white shark species as well as aid in the assessment of the effectiveness of South Africa's conservation measures for white sharks.

This research will also aim to identify the population composition to ascertain what aspect of the population is utilising Mossel Bay. It will also examine the seasonal and spatial patterns in abundance to understand habitat use within the bay, providing insight into the ecology of this species.

Laser photogrammetry is currently being looked at to improve the estimation of the total length of white shark individuals. This method employs a pair of lasers that are adjusted to represent a specific distance. Because the two laser points are of known length, video stills can be used to estimate and determine the size of the shark. Ultimately, this will increase the accuracy and validity of the photo identification process.

There have been other projects on which work has started. Since setting out in 1998, the research that Johnson and Gennari have conducted in the bay was effectively limited to great white sharks, primarily looking at the behaviour as well as using acoustic telemetry to establish the thermal physic-ecology of the white shark. As more

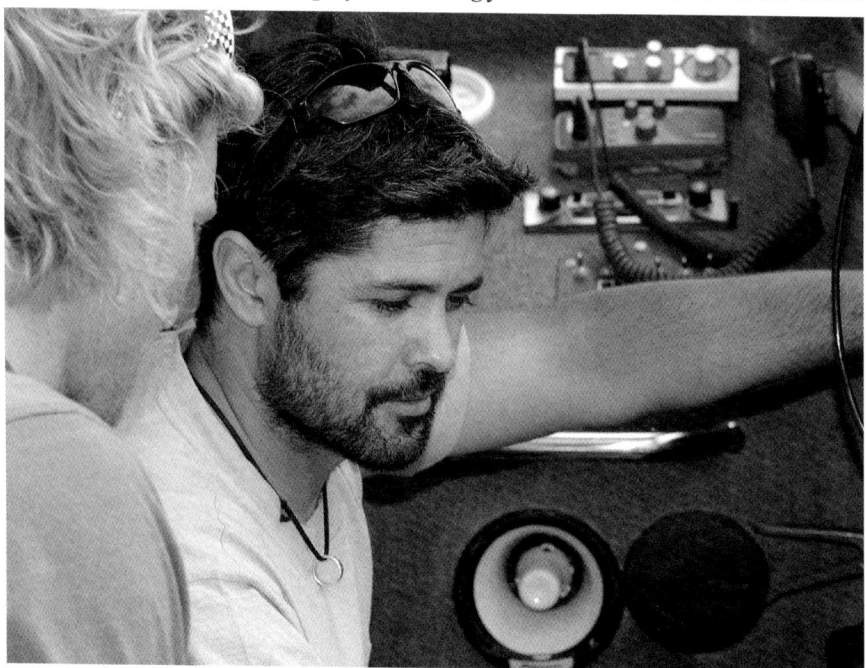

Ryan Johnson and a fellow scientist compare notes while on a research project.
(Photo: Oceans Research)

academics and students have become aware of Oceans Research, research at the institute has diversified.

The body has begun a variety of oceanographic projects to monitor water conditions as well as oceanographic conditions within the bay, which in turn tie into the behaviour of the bigger animals such as the great white shark, dolphins and whales. While standing on its own as an important study – particularly in the development of various industries that could affect water quality in the bay – Oceans Research has also extended into marine mammal research.

The Mossel Bay Cetacean Project is currently run by Oceans Research and the Mammal Research Institute of the University of Pretoria and involves working to gather baseline data on population studies and movement patterns. The project is also looking at how the human development impact affects the behaviour patterns of cetaceans, as well comparing this data with behaviour patterns of cetaceans in a bay that actually has very limited human development present (such as Vales Bay, a site commonly used by the threatened humpback dolphins).

This project focuses on three dolphin and three whale species commonly found in Western Cape waters: the bottlenose dolphin (*Torsions adjuncts*), humpback dolphin (*Sousa chinensis*), and the long-beaked common dolphin (*Delphinus capensis*), as well as the humpback whale (*Megaptera novaeangliae*), Brydes whale (*Balaenoptera edeni*) and the southern right whale (*Eubalaena australis*). At the time of going to press, this project had been running successfully for more than a year and future data will help to identify spatial and temporal habitat-use patterns, including areas of critical importance.

The project will also look at natural variations in the behaviour patterns and the interaction of these species. This will create an information baseline against which to measure the potential impacts on these animals of future human activities such as fish farms and the expansion of the Petro-SA oil refinery infrastructure in the region.

Finally, the information will be used to identify appropriate recommendations to alleviate any negative effects of these growing industries.

In 2009, Oceans Research acquired ownership of the Mossel Bay shark and ray aquarium. The facility was able to attract an incredibly talented aquarist and marine biologist from London, Adam Johnstone, a graduate of Kingston University and recipient of the United Kingdom Tetra Award.

Johnstone left a large well-established aquarium in London to help reconstruct and develop the Mossel Bay aquarium. His draw was the appeal that half of the aquarium's facility would be dedicated to conducting research projects, coupled to Johnstone sharing Johnson's vision that, while the aquarium is an important display for public education, the research conducted on the aquarium species is as equally important.

Johnstone is currently working to further develop the Shark Lab aquarium and strengthen its identity; as well as its role in conservation and education.

The aquarium projects investigate several biological aspects of small sharks, specifically *Scyliorhinidae* and *Triakidae*, which live in the coastal waters surrounding the facilities.

The Shark Lab is an active participant in the ORI (Oceanographic Research Institute) tag and release scheme.

As part of the programme, the Shark Lab aims to establish baseline data including relative abundance, species and size composition, growth rates, sex ratios, habitat utilisation and preference, spatial and temporal distribution, seasonal variation, residency and site fidelity, individual recognition, survival, mortality and recruitment, recapture rates and potential shifts in these following developments within the bay.

This tagging project carries out a regular and consistent catch-and-release programme at reefs in and around Mossel Bay by the collection of sharks from free diving, traps, as well as rods and reels. This long-term tagging programme will strive to determine recapture and mortality rates, time at liberty/longevity, movements, migrations, seasonal distribution and dispersal rates, age validation and growth parameters for benthic sharks released into the bay.

It will also assess tag retention/shedding rates, longevity and time at liberty, subject to recapturing and identifying known individuals through photographic identification of spot patterns as

New Zealand-born Ryan Johnson has made South Africa his home and underwater lenswoman Fiona Ayerst, his partner. Almost single-handed, to start with, he put Oceans Research on the map. (Photo: Fiona Ayerst)

well as utilise detailed morphometrics and length/weight ratios to assess fitness upon capture, post release, and on recapture to assess the effects of captive environments on the fitness of sharks.

The Shark Lab is also working to create baseline research on tonic immobility patterns in the different species and sexes of the benthic sharks in the aquarium. Tonic immobility, also known as death feigning (which occurs in many species throughout the animal kingdom) is still a phenomenon that lacks complete understanding.

TI is known to occur in five classes of invertebrate and all vertebrate species, excluding jawless fish (superclass: agnatha). (Whitman, 1984)

The TI-response has been observed in a wide variety of cartilaginous fish, including lesser-spot dog fish, spiny dogfish, smooth dogfish, tiger shark, sandbar shark, lemon shark, blacktip reef shark, whitetip reef shark, caribbean reef shark, leopard shark, swellshark, california round ray, cownose ray, clearnose skate, southern stingray, shovelnose guitarfish and thornback ray[1]. These experiments require artificially putting the aquarium species of shark into tonic by manually inverting them horizontally.

The record of induction time into tonic, duration time of tonic, and its success will determine if these factors are independent of different species and sex for various endemic sharks. The data acquired from this research is extremely valuable as it will facilitate the expansion of a greater understanding of evolutionary significance of this response in elasmobranchs. Additionally, this baseline study will aid in establishing aquarium protocols when conducting examination, husbandry and medical procedures.

Feed regimes are also being analysed. For apex predators, studies of food rations, food retention times, and food passage rates are an important component in many ecological studies.

The nature of such investigations largely necessitates research within captive environments. Such experiments are essential for bioenergetic studies. By adopting targeted feeding, in combination with detailed record-keeping, estimates of daily ration and subsequent growth rates can be relatively easily monitored. In addition, monitoring voluntary feeding frequency, following satiation, can be adopted as a proxy for gastric evacuation time. Observing the time between feeding events over a range of artificially manipulated temperatures can be used to observe how temperature affects transit time in cat sharks based on the Q10 temperature coefficient principle.

These studies are imperative, not only to these specific shark species but also to establish general methodology and experiment procedures to apply to larger shark species in less controlled environments.

Moving forward, as the work done by Oceans Research matures, Johnson wishes to tackle important issues concerning sharks and other important megafauna, while also addressing projects which the public feels might be important.

Ambitiously, he is eager to address issues that affect the relationship between sharks and humans, by looking at how white sharks utilise the bay and interact with humans. He hopes to try to determine methods which might mitigate the impact of shark attacks. Oceans Research also plans to embark on many projects in the future researching other shark species in the bay.

Through observation, marine scientists are suggesting that Mossel Bay is a nursery for hammerhead sharks, as well as host to a resident population of raggedtooth sharks, which Johnson is determined to study.

In addition, pathology labs have expressed interest in collaborating with the facility, the idea being to branch out further into the technical and physiological compositions of the white shark. The research will

Shark egg attached to kelp. In Cape waters, these are constantly washing up with the tide.

examine heavy metals such as mercury in shark tissues, specifically shark fins which some humans want to consume, diseases in sharks and other marine megafauna, as well as sampling bacteria to examine if sharks possess drug-resistant blood cells.

As Johnson admits, he is in the right place to do all that research, since South Africa boasts – without question – the best and most successful shark resources in the world for study purposes.

The coastal species of South Africa are relatively healthy due to a very active government who has put a lot of protective legislation in place in an attempt to protect sharks and to sustainably manage them. Moving forward, we need to continue to attract collaborations and partnerships with international scientific groups and marine conservation organisations, as well as the academic community while we strive to take on other remarkably interesting research projects.

Oceans Research's ultimate goals are as ambitious as they are vast. The organisation is keen to contribute their research to help develop and solidify South Africa's place in the world of marine research and conservation. In this, the role of the great white shark is seminal: it is an iconic species that has captured not only the imagination of the public, but also the imagination of a vast scientific community.

Sijmon de Waal took this unusual photo of a spotted wobbegong (*Orectolobus maculatus*) while visiting his sister in Australia in 2011.

For many years South Africa's white shark reputation was limited to marine ecotourism, such as cage diving. With ecotourism in South Africa thriving and the research on white sharks still in its development stage, Oceans wants to be at the forefront of world research on the white shark. Johnson feels South Africa deserves to be at the forefront of marine academia and hopes the research being done at the Oceans facility will promote South Africa's reputation as an academic hotbed of marine research.

It is important as the country strives toward becoming a leader in social and economic development that it remains at the head of marine conservation. Johnson hopes Oceans Research's constant advancements and scientific findings will aid South Africa in this fight, while also cementing its place at the top of marine research and conservation.

1 Whitman et al, 1986; Watsky & Gruber, 1990; Davie et al, 1993; Henningsen, 1994; Holland et al, 1999; Heithaus et al, 2002), refer to Whitman et al (1986) and Henningsen (1994).

Sijmon de Waal

CHAPTER SEVENTEEN

ANDRÉ HARTMAN – TRIBUTE TO A GREAT SHARK DIVER

Long before the *Discovery Channel* launched its biannual *Shark Week* series, André Hartman was already recognised as one of the leading authorities on free-diving with sharks. He'd been doing so for many years, first as a youngster in Cape waters and later as a member of the South African National Spearfishing Championship Squad. But even then, he was regarded by his buddies as being somewhat different from the rest of the gang.

André Hartman is one of those rare modern-day adventurers who seems always to have had little regard for conventional norms. A huge, bearded man, he's to be found barefooted each morning when the shark viewing boats set out from tiny Kleinbaai Harbour. In fact, I can't recall having ever seen him wearing shoes, or even a pair of flip-flops.

Not wearing them has become something of a trademark with this man who started diving when his father bought him a speargun at age 14. He made the Springbok spearfishing team when he was 23 and his first international competition was against the Americans at Plettenberg Bay, which he helped South Africa to win.

It's André's predilection for the unusual that seems to have caught the imagination of most of his clients. Word has it that the last time he transited Heathrow Airport on his way to make a shark film with *Discovery Channel* off Guadeloupe, he arrived barefooted at the busiest airport in Europe. When I visited him at his Kleinbaai home in the spring of 2007, he denied it and frankly, knowing the man the way I do, and over several decades, I tend to believe him.

For all that, André Hartman is a most unusual fellow. He has some remarkable insights into the way that great whites – and all the other sharks in South African waters – live, hunt, feed, procreate, squabble among themselves, react to divers in the water, relate to other marine life and die. He is also one of a tiny handful of active divers who has gone into the water among great white sharks outside a cage.

At the same time, he'll tell you that it is not something 'that I just do'. The situation is carefully monitored beforehand, he likes to stress. Moreover, he always ensures that there is an 'escape route', if things start to get 'aggressive'.

In one of the *Discovery* films with which he was involved, he spent a good deal of time in the water beyond the protection of a steel cage with a female diving assistant. It eventually got to the point where one of the more aggressive predators cut the pair off from the boat. That meant that they didn't have access to their cage, since it was tethered alongside.

Somehow he managed, but it was tight because they'd become prospective prey to a particularly aggressive brute. It took a little while for André to shepherd his charge back to safety, and only then did he leave the water.

André's *pièce de résistance* has always been to ride the dorsal fin of a great white, which he has done numerous times.

As he'll tell you, if you're contemplating the move yourself, 'don't stay attached too long because the shark then swims away from the boat, and the boat is your safety net. And then, if one of these brutes gets aggressive, you'd still need to get back on board,' he warns.

He's also gripped the tail of a great white while underwater, but maintains that that's more difficult than the dorsal because the shark is constantly on the move. 'These sharks don't appreciate that kind of resistance to their mobility,' he argues.

It is not surprising therefore that André has played host to a variety of international figures including Peter Benchley, as well as the renowned photographer David Doublet of Washington's *National Geographic*, whose offerings has been seen by generations of underwater enthusiasts. Another visitor to his pad at Kleinbaai was Steve Irwin, Australia's late, great Crocodile Hunter.

'Steve dived with me and actually managed to touch some of the bigger sharks. But he wouldn't leave the cage … said afterwards that he does one thing and I do another and that he didn't want to piss on my fire …' André's sense of humour usually elicits a huge guffaw when he's onto something that tickles his fancy.

A familiar face during America's biannual *Shark Week* – when dozens of the best films involving sharks are screened and which has the highest viewer ratings – André Hartman has worked on many programmes for a variety of international film companies in some of the remotest corners of the globe, including Dyer Island. Other trips have taken him to Mexico's Baja Peninsula, Guadeloupe and a variety of California islands. He has also appeared on Britain's Independent Television Network as well as on *Blue Planet*.

More recently, a German company gave him the prototype of a small two-seater submersible that they suggested he use among the great whites off Dyer. Open to the water, with only a row of metal bars protecting the diver-pilot and his passenger on board – or rather, in-board – they need to wear scuba gear in order to submerge. The machine is propelled by a pair of electric engines that allow for a steady six or eight knots if there's no current. I took a photo of the device and elicited promises from André to accompany him the next time he went down.

Like others of his ilk, André Hartman has had many moments in the water with aggressive sharks, some that he remembers better than others.

'My first aggressive shark was at Partridge Point in False Bay. I was there with a buddy and we saw a lot of seals on the rocks, but curiously, not a single one in the water. We thought that a bit unusual, but it didn't keep us from diving. Obviously, if it happened today, I wouldn't be near the place.

'So we started looking around for fish to shoot. Finally, my buddy said that he didn't feel good about the place and was going ashore. I kept on diving and actually stayed out another two hours before calling it a day.

'I finally got close to the kelp with 13 fishes on my line – I remember it well, because it was supposed to be an unlucky number. I towed my float a short distance behind me and just before I made the final push to land, I looked back and saw this massive great white directly behind me. The monster was obviously onto me.

'The strange thing was, that the very thought that went through my mind was that this was really a magnificent creature. It was beautiful, one of God's great creations and, frankly, that's still the way I regard these beasts. The prospect of any kind of threat didn't enter my mind. Not yet, anyway.

'Things quickly changed when it rushed me, which was seconds later. I pulled my gun up and aimed at the shark's head. That was when it opened its jaws and clamped them shut around my speargun, which was already halfway down its throat.

'The forward momentum of the shark coming at me, thrust me

Sijmon de Waal

backwards through the water, several metres in fact. It continued chewing on my gun and then released it, making a right mess of it too. Cost me a bit of money as well, as I had to get a new one.

'Moments later it attacked a second time. Again I used my gun to fend it off. Obviously all this was taking place at pretty close quarters; in fact, there were moments when our heads were almost touching. Finally, I tried sticking a finger in its eye, and that might have been the turning point, because the shark kind of spat out what was left of my gun and swam away.

'I finally got ashore as if I'd walked on the water. It wasn't a long swim, but I reckon I must have broken all records.'

The next five years saw André Hartman involved with sharks on several continents. As he says, he saw a lot and he shot a lot.

'By the time I got to Struisbaai, some of my old spearo buddies had turned to wreck hunting. I also tried my hand at it and there were times when we did quite well.

While at Struisbaai, André dived the HMS *Birkenhead* and was always intrigued by the number of bronzies lurking off Danger Point.

'There was also a resident great white: a big bugger of four or four and a half metres … always there, in the vicinity of the wreck.' Some of the divers even took photos of it, he recalls today. But that was one of many great whites in the area, with Dyer Island barely 20 minutes away by boat, he'll tell you.

Great whites, stresses André, are not the only problem in these waters. He was diving once off Struisbaai when his buddy shot a sizeable yellowtail. He strung it onto his line – which was attached to a buoy – and then went on with the hunt. By then André was shooting fish for himself. At one stage he turned around to see the yellowtail that his buddy had previously shot, coming directly at him. It was still very much alive, even though it was still connected to the line on which it had been strung.

'There it was, rushing at me at a great pace with a large bronze whaler following behind with its jaws open. It didn't help that the fish swam through my extended arms and that I had to use my speargun to fend off the shark, which was partially successful, though it ripped my rubber slings apart and that was more outlay as I had to replace them both.'

He has his own views about the astonishing number of great white sharks in the area around Dyer Island. It has always been like that, he remonstrates, countering arguments that it's been tour boats that have brought them closer inshore.

'It was that way during those early days as well, before the area became commercially exploitable. I was with one of the first French film makers who came here and we anchored on the northwestern shore of Dyer Island and went overboard, but without putting out any bait.

'I immediately spotted a number of sharks, which was when I called the guy and he joined me. Within two hours, we saw about 30 great whites – 16 in the morning and almost as many that same afternoon. They didn't hassle us and weren't unduly inquisitive.'

Right now, there are even more, he reckons and, of course, there are many more boats. But as always, a lot depends on the weather: big swells and bad visibility can affect the situation. Then you don't see that many.

One of the interesting sharks that André Hartman kept tabs on over the years was a hunchback great white that regularly visited the boats that emerged from Kleinbaai Harbour.

'We called her Quasimodo, a female that had obviously had her back broken at some stage. It seemed that her spine might have been almost severed by another shark, because you could see the original bite marks on her back.

'This impediment caused the predator to be almost immobilised. If she was a human, she'd be a cripple. In any event, this old girl was slow and cumbersome and we spotted her often, which was why we were aware that she almost certainly couldn't hunt down seals. She couldn't even turn to take the bait from our boat if you pulled it sideways. Yet, she stayed sleek and fat, so she must have managed somehow. We observed her in this area for about three years. Then she just disappeared.'

André Hartman summarises: 'What one needs to accept about great whites is that they are all so very different. Each shark has a personality of its own. Some are aggressive and will come onto the bait immediately. Others will hang back until the area is clear of other predators. I saw one shark chewing away at the bait and a large male come up fast from below and bite it severely in the gut. It was a serious effort, because the shark's entrails were exposed. Then one starts to wonder whether it will survive such a wound.

'There is also the eternal question of intelligence which, when you've worked with these animals long enough, you only begin to get an inkling of.

'Some sharks are clever: they know what they want and they also know how to assert their presence. There are various levels of aggression too, which is testified by the bite marks from earlier attacks, all of which are visible. Others are plain stupid: they react badly under stress. But then that's a bit like us humans …

'There is a very distinct pecking order within the shark community. You see it when you have a small pack that arrives around the boat: one shark will clearly be the dominant Alpha Male and the others will hang back until that one is done. They like to eat by themselves and not in groups,' was André Hartman's final word on the subject during that visit.

CHAPTER EIGHTEEN

FIONA AYERST – UNDERSEA LENSWOMAN EXTRAORDINAIRE

Why sharks? This is a question I am repeatedly asked, especially from non-divers. My retort is usually a perfunctory: 'Why *not* sharks?'

Sharks are streamlined, muscular and nature's perfect creation of something unimaginably beautiful. Those of us who have dived among them are always in awe of the effortless way they are able to slice through the water.

Some of these creatures have enormous reserves of power, witnessed by the way great white sharks go into the attack, usually from the bottom up, as they do when targeting seals. By the time a large 'white' reaches the surface and strikes, it has attained an astonishing speed. Which raises the question: having a model like this to work with, must surely be every photographer's dream.

Yet, the concept of shark attack, invariably sensationalised by the media, chills. Most humans tend to have a reasonably well-founded fear of sharks – accentuated each time there is another attack by newspaper reports that often ignore much bigger stories. The truth is that the majority of people simply cannot deal with the idea of somebody being eaten alive.

That aside, it has been my personal choice as an underwater photographer to photograph sharks in their natural environment. I have done so for many years and have tens of thousands of images to attest to my enthusiasm, not yet dampened. That said, I guess I could call myself a specialist shark photographer.

I actually enjoy the challenge of being in the water with these creatures and taking the kind of photographs that I would like to regard as creative or possibly even artistic. I relish the challenge of bringing back from the deep fine images of sharks in their natural environment. My perception can hardly be equated to regarding these beautiful predators as vicious, all-consuming flesh-rippers, constantly in search of humans to attack.

In fact, if you examine available statistics, you will soon conclude that, relevant to the millions of people who are in the sea at any one time – worldwide – there are astonishingly few shark attacks. The recorded instances of shark attack number perhaps a few dozen a year, which just about says it all. It is reckoned by those who compile these statistics that something like 70 million sharks – give or take 10 million – are killed by humans each year. You do the maths!

What is important is that it is necessary for people to accept that diving with sharks is a relatively safe pursuit. There are exceptions, of course, such as diving in murky estuaries or ignoring certain definable threat signals that indicate a shark that might not be comfortable with an immediate human presence (such as great white sharks when feeding), but these are all matters that can be countered by experience and listening to others who have been there …

Also, it helps to dive with individuals who are experienced in such matters, know what they are doing while underwater and are well versed in shark behaviour. It is obvious that the most experienced 'shark divers' – people like Walter Bernardis, Ryan Johnson, Mike Rutzen, Tommy Botha and quite a few other notables – have hundreds of hours in the water with sharks.

Speak to these individuals and not one would admit to larger sharks being anything but predators and, as a consequence, potentially dangerous. They are likely to tell you that sharks might be compared to many species of wild animals ashore – lions, leopards, tigers, bears and other animals that hunt in order to survive. You simply would not get out of your car in a game reserve, for instance, and stroll towards a pride of lions without somebody at your side that has good experience in such matters and is, in all probability, armed.

The difference with sharks is that when you are in the water with them, you don't need weapons with which to protect yourself. Most times you can swim into their territory, move about for an hour or so and emerge at the end of the dive totally awed, but unscathed. Most novices who swim and dive with sharks for the first time admit afterwards that they viewed these beautiful creatures in a totally different light after that first extremely exhilarating, mind-blowing experience.

Try it for yourself, but as with all other kinds of underwater activity some distance from the shore, never dive alone. Approach one of the many well-established dive companies that have proven safety records and speak to their divers. Bottom line: establish the parameters, and only then do you go for it, but first you need to do your homework, because there are some unscrupulous operators out there …

The second question often raised is usually something about my 'favourite' shark. This is far more difficult, as most times it depends on where I have been and what interactions I have had with these predators at the time. Most species have been favourites at one stage or another and that range includes blue sharks, tiger sharks, makos, great whites, whale sharks, bull sharks and even smaller species like pyjama or the appropriately-named shy sharks.

Indeed, I don't believe there can be an *overall* winner, but if pushed, I guess it would have to be tiger sharks because of their graceful beauty and the nonchalant manner in which they approach every situation and then wind up taking complete control. Wolfgang Leander, that wonderful 'Old Man of the Sea' who, well into his seventies and still diving, shares quite a few of these sentiments.

I have photographed tiger sharks along the Bahamas Banks in the Western Atlantic and also in the Scottburgh area off the coast of KwaZulu-Natal in South Africa, but my most exciting experience of these wonderful creatures simply has to be diving with them in the Bahamas at *night*!

Fiona Ayerst

I found that there was little change in their countenance from daylight hours to night, such as I have experienced with other sharks, including the raggedtooth or sand tiger sharks. My most scary experience with tiger sharks was when they dummy-rushed at me from below, which underscored the reality that tiger sharks do sometimes have a nasty habit of zooming straight up at you from somewhere in the depths.

In the Bahamas you dive with tiger sharks at around five metres on the sand, so conditions are usually perfect. In wilder, southern Indian Ocean waters, the opposite is true. With a chum ball in place, you are generally suspended in the sea at about five metres, customarily with five to ten metres of open water below you.

There have been dives at South African locations where I have had no option but to push the dome port of my camera housing right onto the oncoming snout of an over-enthusiastic tiger shark. Their noses do look a little like marshmallows as they come at you at speed, but we are all aware that there is an impressive array of teeth behind that deceptive veneer. Though it has not happened to me – yet – tiger sharks are known to sometimes lock their jaws onto underwater equipment and, occasionally, even make off with camera equipment. It then becomes a little expensive ...

There are more of my dive companions who have a predilection for great white sharks, though there are precious few enthusiasts who have actually free-dived with them without the protection of a steel cage. A large part of it stems from the enormous power of the white, its uncompromisingly lethal-like appearance and, let's face it, those awesome jaws, which sometimes defy description.

I have done some free-diving with these sharks while I was involved in making documentary films in the Gansbaai area. On one occasion the visibility was excellent, at about eight metres, and one of our crew members dropped a metal pole that had been baited with a tuna head into the sea from the boat. I offered to retrieve the pole, but on reaching the bottom, I couldn't find it.

Because I was on scuba, I started swimming around in my search when I suddenly sensed something large move in the water over me and from behind. I guess it was the large moving shadow that alerted me, because when I looked up, I was faced with a 3.5-metre great white shark that had already started to circle. Though the shark was not overtly threatening, its size alone was pretty damn intimidating and I confess to a sense of sheer terror.

Suddenly the shark dropped down in the water a metre or two

towards the reef below and then swam straight towards me: it was coming in at quite a pace. Aware that I did not dare to act like prey and try to flee, I simply extended myself lengthwise in the water to try to make myself look as forbidding as possible (fat chance of that) and at the same time, screaming and madly blowing bubbles.

It was a measured approach because I actually wanted to attract the predator towards my head rather than my more exposed legs. If I could achieve that much, I was confident that I would be able to duck under it as it swam over me. To cut a long and rather frightening story short, exactly this happened seven or eight times. In the meantime, I also made use of some young kelp plants that were growing near my feet to pull myself down fast each time the shark came close.

For a while, I sensed that if this situation was allowed to continue, it would otherwise have been construed as comical, except that there was an increased threat from the shark with each circle. Luckily a second, smaller white shark appeared shortly afterwards and its presence broke the rhythm of the circling shark because it ended up chasing the little critter away. About then I spotted a patch of kelp nearby and moved in that direction to hide.

To my chagrin, the great white wasn't yet done with me. It returned once more and again took to circling the kelp. At this stage I was starting to get a little worried because my air supply was down to about 60 bars. Then good fortune played a hand, because after I'd ventured towards the rear of the kelp bed for the umpteenth time, I found the metal pole. I wasted no time in removing the tuna head, which I threw in the general direction of the shark, surfacing at the same time.

I wasn't able to locate the boat from the bottom, which meant that after I got to the surface it was still 50 metres away, which meant still more lonely minutes in the water with a large and seemingly hungry shark. Because I really didn't want to head down to the bottom again, I made the decision to remain on the surface, but with the pole sticking straight down towards the sand from my mid-region.

At the end of it, I didn't see that shark again. I swam back to the boat. Nowadays if I am diving with great whites, I always try to take along something hard and stick-like which I can use to push these creatures away. The professionals used to call them 'billys' – metre-long broomsticks that have sharp nails embedded into their business ends.

To this day I have never again been threatened, but there is some comfort in knowing that I have something to hand which I could use should the need arise. A diver's pliable hand is hardly adequate to push away a great white: it could be mistaken for food, especially if it suggests signals of distress.

Something hard and inert, like a small stick, is the way to go.

I have spent many hours photographing the always-gorgeous bull or zambezi sharks around southern Mozambique.

On one trip my partner, marine scientist Ryan Johnson, and I decided to combine a photo safari with a healthy dollop of scientific research. The idea was to tag a bull shark so that scientists could follow its movements and learn more about these fascinating and much-maligned animals. However, we were faced with a dilemma: how do we tag a bull shark?

We were both firmly opposed to use line and hook to catch the shark, if only because the stress potential would have been severe. Instead, Ryan decided that perhaps the best way forward would be to try the Hawaiian-sling method and I was promptly delegated to film the tagging procedure. At the same time, both of us were certain how a bull shark might react to being shot at with a sharp dart: we were both sure that it wouldn't be too happy.

It wasn't long before a perfect opportunity presented itself when an apparently fearless shark arrived on the scene. Ryan wasted little time: he took aim and fired, striking the shark squarely towards the bottom of the dorsal fin. I was in a perfect position for the tag. However, never having tagged a bull shark before, it seems that

Fiona Ayerst

we underestimated either the thickness or the tensile nature of the shark's rough skin, which meant that the tag dart just popped straight off and the predator took a fast exit straight for the distant blue.

Though we were initially relieved that the shark had disappeared as quickly as it did, we were faced only moments later with the predator heading straight back in our direction, almost like a missile. Without warning, our original quarry was making a beeline straight towards his now-identified adversary, Ryan. Fortunately, he still had the sling pole in his hand, but the shark didn't come close enough for Ryan to have to use it. What was obvious to us both was that this was one extremely angry shark, and it wanted us to know it.

Curiously, a few minutes later it was swimming around again as if nothing had happened.

Seven-gill cow sharks, prolific in the Western Cape, are among the more interesting sharks with which to dive.

With their broad, blunt noses and large crocodile-like eyes, mature specimens look almost like prehensile dinosaurs. And here too, first impressions can be deceptive: they appear to be so relaxed, and it stays that way until you take anything electronic into the water. I have seen their behaviour change dramatically when exposed to the electronic squeal from camera strobes and even an underwater voice recorder. Their predilection is to nose and nibble on anything that emits a noise or a pulse, much more so than other sharks with which I have dived.

While all sharks appear to be interested in strobe lights, cow sharks became so much more animated when these are around.

A diving friend recently embarked on an interesting experiment with this species, the idea being to test a new shark-repellent device. The results were hilarious. Instead of swimming away at the sight of humans as these sharks normally do, they were so attracted – either to the noise or vibrations emanating from the repellent – that they were all over the divers: some even having to have their attentions physically fought off. Their actions were not aggressive in any way, but the repellent seemed to have just the opposite effect: the sharks became distinctly amorous. Photographers often experience something similar, due probably to the electronic devices they take underwater.

One of the great experiences in any diver's life is a visit to Bassas da India, a lonely, isolated coral atoll halfway between central

Mozambique and Madagascar. With an enclosed water area of roughly 18 kilometres across, Bassas had always been regarded as something of a notable shark hideout.

We went to Bassas in search of these wonderful creatures a couple of years ago and, sadly, it was an enormous disappointment because there were almost no sharks in the atoll to greet us. Nor were we alone: Mike Rutzen of great white fame, has since taken a BBC film crew to Bassas da India, and though they were in the area for days, they never saw a single adult shark. All had apparently been taken by Asian longline fishermen that are today stripping vast expanses of the Indo-Pacific basin of a resource that is not only vital for the survival of the maritime ecosystem, but once depleted, takes decades to replicate.

I was obviously deeply disturbed about this revelation, and it stayed that way until one evening when Ryan and I found us out on a dive after the sun had already slipped beneath the horizon. As shadows lengthened, so several young Galapagos sharks – they are prevalent there – seemed to nudge in closer to us. I couldn't understand why we were being buzzed, and in such close proximity, so we hot-tailed it back to the boat. That was when I spotted Ryan discarding three small sardines that he had been hiding in one of the pockets of his BC.

Unfortunately, the sharks were already hyped-up by the smell and the fact that we had ditched the food source didn't seem to deter them at all. Nor did the ever-decreasing light help: I felt as if I was swimming through a ghost ride at a fun fair – with shifting shadows rushing at me from all sides and then veering away at the last second. Never have I been so relieved to get my butt back on board a boat after a dive. Fortunately, the sharks we encountered were all juveniles, from about one to 1.5 metres long.

When diving, I always tend to feel more secure when I have my camera with me. This is probably only a psychological 'safety blanket', but it certainly seems to work for me, even though I am aware that this equipment – with its sleek silver casing and enticing electronic sounds – may actually be attracting rather than repelling sharks.

In truth, I have never felt threatened enough by a shark to immediately get out of the water and I have been diving with them for more than 15 years. I think it is important for photographers not only to study, but also to appreciate the behaviour of sharks – both to get better pictures as well as for their own safety.

That said, I personally know at least four photographers who have been bitten by sharks while working with these creatures. In all cases the bites were minor and the consensus was that there had possibly been a case of mistaken identity. It is a simple matter for a feeding shark to confuse a glinting camera for a fish and, as we know, feeding sharks are enormously impulsive. In all cases it seemed as if the sharks were after either cameras or strobes or, even more likely, other fish close by.

Even with these odds, stacked as they are in favour of the photographer, anybody making contact with sharks in their own milieu should be both respectful and circumspect.

So what is it that makes anybody in his or her right mind become a shark photographer? It is quite a mix really, not the least being the relatively easy access to areas where there are still sharks around that the pursuit allows.

Sharks are never easy to encounter in open waters if somebody is not chumming for them. As you will have read elsewhere in this book, these creatures are being systematically and extremely efficiently slaughtered on a colossal scale. Their numbers are in steep decline and should this trend not be checked, there is no question that some species of shark will be seriously threatened. Some already are.

There are areas off our shores where, less than a decade ago, we could dive with a dozen or more bull sharks each time we went down: go there today and you'd be lucky to find one. In recent years I have observed a massive drop-off in numbers of my favourite quarry. I – and many others like me – am only too aware that the clock is ticking for the shark populations of the world.

That said, it is important to dive with like-minded people and with dive operators or organisers who know exactly what they are doing. Unfortunately, sharks generally need to be fed to draw them closer, and that means that most operators use chum to attract them. In one sense, this practice is a double-edged sword and it brings with it a situation that could be potentially dangerous. You are in the water with sharks that are fixated by the food on offer and in such circumstances, mistakes can and do happen.

Sharks have been known to be attracted to the silver flash of cameras, and it seems that some of the smaller glinting cameras may sometimes remind them of the sardines they are often being fed. The same with a diver's hand protruding from a wetsuit: it could easily be confused for a sardine! And that, too, has happened more than once ...

There are operators who are testing alternative attractants as well as methods of attracting sharks that do not actually involve feeding them, but it is a slow process of trial and error. Personally, I hope the sport will evolve to this.

Over and above some of the problems that are linked to access, one also needs to be patient, relatively fearless and passionate, reasonably fit and pretty well off financially to enjoy the sport. Like golf, diving is not cheap and chasing sharks around the globe with an expensive camera set up to work efficiently underwater, is hardly a pastime for the impecunious.

Also, and this is important, shark diving is not something for the fainthearted!

Fiona Ayerst

CHAPTER NINETEEN

SHARK ATTACK: THE WILLIE VAN RENSBURG STORY

Former Argus *journalist and an acknowledged crayfish poacher, Jim Pennrith has done more diving in Cape waters than most. Of necessity, he became an authority on shark attack. He spoke to one veteran Cape diver who survived a shark attack.*

F ather's Day, 19 June 1988, dawned like any other day for Willie van Rensburg, then a 36-year-old building contractor in the picturesque resort town of Hermanus, on South Africa's southwest Cape coast.

Although it was a Monday, a working day, he decided to take the day off. He decided that he would spend it skin-diving and spearfishing with his son Willie and his son's friend Wilhelm, both keen 16-year-old divers. Van Rensburg swept the sea below his home with his binoculars; as he recalls today, the water looked perfect for snorkel diving, clear and calm. He would take his ski-boat out to his favourite diving spot, a place known as Skulphoek ('Shell Corner'), some three kilometres from Hermanus' new harbour in the direction of Cape Town.

Soon the little craft was bobbing on the blue, anchored in eight fathoms of water, a mere 200 metres offshore on the seaward side of a reef Van Rensburg knew from long experience was a good spot for spearing the tasty pan fish known locally as *galjoen* ('galleon').

Leaving his two youngsters diving near the boat, Van Rensburg finned over to the reef and floated quietly on the surface in his black neoprene wetsuit among the waving fronds of sea bamboo, looking down through the clear water to the sandy bottom some seven metres below. In varying conditions, he had surveyed the same reef through his face mask hundreds of times over previous decades. It was a peaceful scene as he closed his eyes to relax before diving; he had done this countless times before in his many years as a *perlemoen* (abalone) diver and provincial spearo. This time, however, the dive was to be very different.

'I was floating there with my eyes closed when I felt something give me a tremendous bump on my left arm. I thought my son Willie had come up from below and had hit me accidentally,' Van Rensburg remembers. 'I looked around for him but couldn't see anything. Suddenly something didn't feel right. I lifted my arm and all I can remember is seeing red; blood was spurting all over the place. I knew then that a shark had attacked me. I spat out my snorkel and shouted to the boys to get back into the boat.

'I was about 60 metres from the boat at that stage and I was dreading another attack from the shark. I rolled onto my back and swam to the boat with my bleeding arm held up out of the water. The boys grabbed me and pulled me into the boat by my good arm. They were very calm, both having done diving courses and seen plenty of shark attack photographs in books; in a way they were prepared for such an emergency, though obviously alarmed.'

When the boys pulled Van Rensburg into the boat, blood was arcing high into the air from his severed radial artery. They hacked off a length of the boat's painter and used it to tie a tourniquet around this upper limb to staunch the bleeding, but it didn't seem to help. Desperately, young Wilhelm stuck his fingers into the ragged puncture holes in the blood-soaked wetsuit and applied direct pressure on the pumping artery, finally stemming the life-draining flow.

For his part, young Willie started the outboard motor, cut the anchor rope and headed back for the harbour. Less than an hour after the shark had attacked, Van Rensburg was lying on his back in the hospital in Hermanus, by then having lost an enormous amount of blood.

He was hastily patched up without anaesthetic and stabilised for the hour-long ambulance ride to Cape Town, where two doctors operated for three and a half hours to put several hundred stitches into the wounds in Van Rensburg's upper and lower arm.

Hospitalised for ten days, he reckoned afterwards that he was lucky he had been wearing his five millimetre-thick rubber wetsuit at the time.

'It helped to keep the torn muscles and flesh of my arm together and I think it was also the reason the shark didn't follow up its attack. They don't seem to like the taste of rubber.'

Van Rensburg still bears the scars of his close encounter, jagged teeth marks from his wrist to his upper bicep. Experts from the KZN Sharks Board subsequently examined and measured the tell-tale crescent bite pattern and their view was that Van Rensburg had been savaged by a great white at least three metres long. About that the experts were absolutely certain.

The diver himself believes the attack was a case of mistaken identity. 'There was a colony of Cape fur seals at Skulphoek and these, plus the tons of fish offal that local trawlers had been dumping into the sea beyond the new harbour must have attracted sharks from miles around. The great white that attacked me probably thought I was a seal and bit me to make sure.' Fortunately, comments Van Rensburg, the shark was only tasting, or, in the lingo 'mouthing'.

'If it had really chomped me the predator would almost certainly have taken my arm right off and, frankly, I wouldn't be around to talk about it.'

These days, now into his sixties, Van Rensburg still dives, but confesses to an uneasy feeling whenever he's in the water anywhere around Hermanus. But then he jokes: 'As they say, you can't live forever.'

There was no such case of mistaken identity when a 300 kilogram raggedtooth shark attacked spearfisherman Manfred Aistrich in the warm waters of the Indian Ocean off the KwaZulu-Natal coast.

The enraged raggie turned on the German diver after he had speared it. He'd tried for a spine shot, just ahead of the dorsal fin in an attempt to immobilise it.

At the time, Aistrich was diving with his spearo buddy John Hughes, one of South Africa's most experienced divers who had attained Springbok colours as spearo numerous times. The two men were on a shark-hunting expedition to try out a new underwater camera. Aistrich subsequently settled in Canada, far from the scene of his encounter with the raggie but, as he has since told friends, including his old dive buddy Jim Penrith, that day remains indelibly etched in his memory:

'We left my future wife Sheila in the boat, which I'd anchored on an offshore reef while we scouted the area for likely sharks,' recalled Aistrich. 'We encountered a pack of raggies close to the reef drop-off. John had the camera while I had a big wooden speargun, with twin rubbers and a drop-off head. I'd shot sharks before and my intention was to try out a new technique.'

Finding the right spot on the shark's head, in a bid to aim for the

Sijmon de Waal

tiny brain is not an easy task, he admitted afterwards. 'But you can also immobilise a shark by breaking its spine and, without the use of its tail, the shark has had it. I decided to use the dorsal fin as the marker and aim as far forward of the fin as possible to hit that vital spot.

'A big raggie was cruising below me and I speared it from above in about ten metres of water; I could immediately see that my spear went right into the spinal cord. With it sticking out of the creature like an aerial, the shark stopped swimming and I thought it was a goner. There were other raggies buzzing round as I pulled on the spear, which came out of the drop-head and, to my surprise, the shark suddenly went berserk.'

With its jaws wide open, the shark went straight for Aistrich. The next second several jagged rows of teeth had clamped onto his left hand.

'As those jaws closed on me, I thought that was it, my end. Blood poured out of the shark's mouth … my blood, clouding the water. With all the strength I could muster, I used my other hand to push its nose away in a desperate effort to free my other hand, but without success.

'The shark then opened its jaws to get a better grip of me, which was when I quickly pulled out my hand. At the same time, I shoved my speargun down its throat, hard!

'The raggie then went into another frantic mode and began biting into the gun just inches ahead of my hand. Probably because it didn't like the taste, the shark spat it out. The next thing it was biting on the spear which it bent almost double on both sides of its massive jaws. Then it broke free and, violently shaking its head, it headed for deeper water.' The shark was still attached to the cord linked to Aistrich's speargun.

By this time John Hughes had called Sheila to bring the boat closer. The two spearos clambered aboard and started to tow the shark closer to shore, until they were just outside the breakers. Aistrich takes up the story again:

'Meantime, I bound up my hand in Sheila's shirt and jumped in to steer the shark through the white backwash. It was still alive and I had an uneasy feeling dragging it ashore. The spear had not broken the raggie's spine and it still jumped about when finally beached. At that point I blacked out ... shock and loss of blood had set in.'

John Hughes had a ringside seat to the unfolding drama from the moment his buddy speared the raggedtooth shark until he finally dragged it ashore through the surf and collapsed. Hughes's recollections of the event are chilling, especially the battle with the shark.

'Once Manfred pulled his spear out of the shark – it was quite a big one and was obviously a bit stunned from the hit – it quickly recovered and first grabbed a fold of his wetsuit. Then it went for him the second time and I thought it had bitten him in the stomach. What followed was something that I'll never forget: Manfred's left hand had disappeared into the shark's jaws and he was putting up a pretty stiff fight to pull it free,' Hughes recalls.

'He seemed to struggle on for an unusually long time, with neither of them prepared to give up the fight. The next thing we were safely on board and heading for the shore with the shark in tow. It was all pretty rough and tumble, but we were young, fit and strong and when Manfred keeled over on the beach, I revived him with a bucket of seawater. That was when I put his mangled hand in a tin plate and, against my better judgement, used our last bottle of cane spirit as an emergency disinfectant. I was pretty reluctant to do that at first, because it was a Sunday and all the liquor stores were shut, but in the end, looking at the terror in Sheila's eyes, I thought I'd better.

'We improvised a tourniquet and kept the bleeding down during the several hours it took us to drive him to the nearest hospital.'

The giant raggie had one surprise left for the shaken trio after it was safely high and dry on the sand. When they cut it open as the final *coup de grâce*, they discovered that it was a pregnant female, full of live, glassy little sharks.

'As these little predators were released into the sea, the one ungrateful juvenile bit Sheila's hand,' said John.

Other shark encounters in southern African waters have had much more horrific finales.

In Cape waters, we had the fatal attack by a great white shark on Durbanville fifth-year medical student Henri Murray in June 2005, while he was spearfishing late afternoon off Miller's Point with his buddy Piet van Niekerk. The shark came for Murray and he managed to evade it twice. Van Niekerk meantime fired a spear into the monster, which was reckoned to be about five metres long.

When he was finally taken, according to Grant Munro who saw the incident from his bungalow, the shark lifted Murray clean out of the water and then disappeared. It was the second attack in months and it literally 'came from nowhere' as his partner later described the event.

Although global statistics over the years have shown that South Africa averages about half the annual attacks experienced in the United States and less than a third of those reported in Australia, shark encounters everywhere seem more likely to be increasing, even marginally so as numbers of people are drawn to the sea for recreational activities. Swimmers, surfers, spearfishermen and even scuba divers can be at risk, though of the hundred or so species of shark that we find in our waters, only half a dozen have been implicated in unprovoked attacks on humans.

Shark experts have no doubt that most attacks in the colder waters of the Cape are the work of great whites (*Carcharodon carcharias*). Those in the warmer Indian Ocean along the Natal coast are usually the result of contacts with raggedtooth sharks (*Odontaspis taurus*) or, more likely, the truculent zambezi or bull shark (*Carcharhinus leucas*), the only species which can be found in fresh water many miles from the sea.

Also on the list of dangerous southern predators are the hammerhead (*Sphyrna zygaena*), the blacktip (*Carcharhinus limbatus*), and the tiger shark (*Galeocerdo cuvier*). Divers used to believe that tiger sharks in southern African waters were a rare event: Walter Bernardis of Umkomaas who dives with them regularly has proved otherwise. There are a lot more 'tigers' about than is generally given credit for.

The tiger shark is often referred to as 'the garbage collector of the deep' by members of the KZN Sharks Board who bring specimens caught in their nets back to their labs for dissection and examination. There is good reason: in the stomach of one four-metre tiger snagged off Durban they found the head and front legs of a crocodile, the hind leg of a sheep, three seagulls, a cigarette tin, together with two unopened tins of green peas: quite a haul!

Statistics also tell an interesting story: the largest tiger shark ever measured was 5.5 metres long and weighed in at 1,524 kg, which in layman's terms, is more than a ton and a half.

Like complacents who subscribe to the fallacy that lightning never strikes the same place twice, divers incline to the belief that if you survive one shark attack, it is unlikely that you'll ever be attacked again.

Events in the spring of 1989 put paid to this mistaken if comforting belief. Cape Town diver Gerrit van Niekerk (28) was spearfishing with three friends in Smitswinkel Bay, near the toe of Africa, a reef-strewn area where the warm waters of the Indian Ocean quietly merge with the cold currents of the Atlantic. Van Niekerk was a few hundred metres from shore and preparing to dive in a kelp canopy. He was in about fifteen metres of water when he felt a tremendous blow on the chest. The thump tore off his mask and caused him to drop his speargun.

After he'd managed to refit his dangling face mask he observed the water around him start to change colour. Then he realised that blood was seeping out of his torn wetsuit.

He later said his first thought was, 'Oh my God, a shark's bitten me ... I'm going to die.' Van Niekerk started screaming in blind panic but regained his senses as he gradually came to realise that his attackers hadn't followed up its initial strike, which was when he started to make for the shore, constantly looking behind to check that he wasn't being followed. As his feet hit the sand on the surfline, he heaved a sigh of relief and as he told someone later, he thought, 'I'm safe now.'

With blood oozing from his wetsuit, the wounded spearo stumbled for help to a nearby cottage and was soon being treated for shock and stitched up in the False Bay Hospital.

Van Niekerk's injuries were interesting. They showed that the shark that attacked him had done so with a sideways slashing movement of its jaws. It left teeth marks on his lead weights and razored through both his wetsuit and rubber undervest to slice into his chest and stomach. Opinion was that the shark had mistaken him for one of False Bay's plentiful seals, but had aborted its attack when it tasted the neoprene/rubber mix of his wetsuit instead of seal blubber. Van Niekerk was asked whether the event would stop him from further diving. He grinned when answering: 'Not on your life ...'

Six weeks later Van Niekerk disregarded the cardinal rule of all divers: dive alone and you die alone. He went to dive alone off Melkbosstrand, on the chilly Atlantic side of the Cape Peninsula and was sadly never seen again.

A week after his disappearance a mutilated human leg bearing the unmistakable marks of shark bites washed ashore in the area. Members of the helicopter crew who had searched for him – and found no traces whatever – said that on the day Van Niekerk vanished they had spotted great hammerhead sharks cruising in the sea below.

It is well known that sharks pick up the vibrations of speared fish and that their death throes are like a dinner gong for any predator in the vicinity. Hungry sharks also have an extraordinary sense of smell. They can detect a single drop of blood in 25-million parts seawater and, like ocean bloodhounds, are able to hone in on the source. Their other senses are also super-sensitive.

They respond to sounds nearly two kilometres distant and their sensory toolbox enables them to register the weak electrical fields which surround all creatures in the water, including humans, down to an incredible five-billionth of a vol. Taken together, this is a

Morné Hardenberg

formidable array of attributes and makes the shark the most sensitive and best-equipped predator on the planet.

But sharks have a favourite item on their menu. Great whites are drawn almost irresistibly to concentrations of seals, of which there are many concentrations in these southern oceans. Without question, this is their favourite food, and three rocky outposts in Algoa Bay, Mossel Bay, and False Bay – all named Seal Island – are like fast-food takeaways for these roaming predators.

False Bay, in particular, is notorious for the size and abundance of its whites, one so big that it was known locally as the Submarine. Danie Schoeman originally called it that because of its size, and by all accounts, it was truly a whopper – the king of all the whites in the region. Some people talked about it being something like six metres long and that shark is still the subject of lore when divers get together and tell of their adventures.

It's worth mentioning that the Submarine in its day was even spotted by pilots flying over the blue 1,000-square-kilometre expanse of False Bay. It stayed around for years and provided spearos, anglers and fishing boats with some nasty surprises.

One September day, the late, great Western Province Springbok spearfisherman Ian 'Tubby' Gericke – together with his spearo buddies Brian Clark and Arthur Ridge were hunting reef fish with a friend eight kilometres out in the bay off Whittle Rock. The reef there rises to within five metres of the surface in a shallow pinnacle, before dropping away into deep water.

Visibility of about six metres was good for the area. After the four divers had speared a number of the fat fish known locally as red roman, Gericke was readying himself for a deeper dive when he noticed a huge shape rising up below him.

'At first I thought it was a trick of the light reflecting from a patch of reef until it got bigger and bigger. My first thought was that it was a killer whale … it just seemed far too big to be a shark,' he recalls.

'The next thing I'm staring at a truly gigantic shark, its head only a metre from my speargun. It was a brute of at least six metres and easily two tons in weight. The monster could have swallowed us all whole and at that moment I thought this was the end of me. By the time Brian Clark came over, I was spinning round and round in the water, wondering when it was going to attack.' The two divers clambered back into their boat and flopped down in relief.

Meantime Arthur Ridge and his buddy were still in the water. He recalls thinking it odd that Clark – usually the last to call it a day – was already in the boat.

'I went down for a final dive, hitting bottom at around 20 metres. I looked around and had the weird impression that somehow the reef was moving towards me ... then the Submarine swam past me with its jaws hanging open, usually a sign that a shark means business. It was so close I could have reached out and touched him. It seemed an age before his tail passed me, which was when I shot to the surface and scrambled into the boat.

By this time Brian and Tubby were chuckling, but they didn't laugh for long. While busy hauling in the anchor, the Submarine surfaced right alongside their five-metre-long boat, its head near the anchor rope and its tail sticking out way past the outboard motor at the stern.

'It lifted its giant head out of the water and stared at us. It had a blank, black eye that seemed as big as a saucer and we all froze with fright,' recalled Tubby.

'By the time its head went down and the shark dived, the anchor was up, the engine that never started easily, turned over at the first pull, and while the shark was vanishing in a swirl of water, we were already racing for the safety of the shore.'

It is worth mentioning that it was also in the vicinity where divers have encountered the Submarine that Brian Clark's boat was swamped during heavy weather late one night. He and several others had no option but to swim to shore, a very substantial task that took several hours.

'There might have been sharks in the water at the time, but it was dark and we didn't see any ... doesn't mean that we didn't think about that possibility ...'

This was also the scene of a subsequent attack on another Western Province Springbok spearo, Attie Louw, who underwent surgery and was in hospital for six days after his legs were badly lacerated by the glancing strike of a great white. Louw had been in the water with buddy Peter Strydom only a few minutes and hadn't shot any fish when the shark hit him with such force as it passed between his legs that it was seen to break the surface along its entire five-metre length.

On the other side of the bay scuba divers Louis Jordaan and Coen Marais fought off an attacking four-metre great white shark with their bare hands, which was unusual because sharks generally don't attack scuba divers in South African waters. It appeared out of the blue while they were chopping up the sea squirts known as redbait (*Pyura stolonifera*) to feed reef fish in about 15 metres of water.

Fiona Ayerst
p o r t f o l i o

Previous page: A duo of inquisitive bull sharks.
Clockwise from above: Father and son peer into an aquarium. • Lemon and tiger sharks in the Bahamas. • Night diving with sharks in the glow from the ship's light.

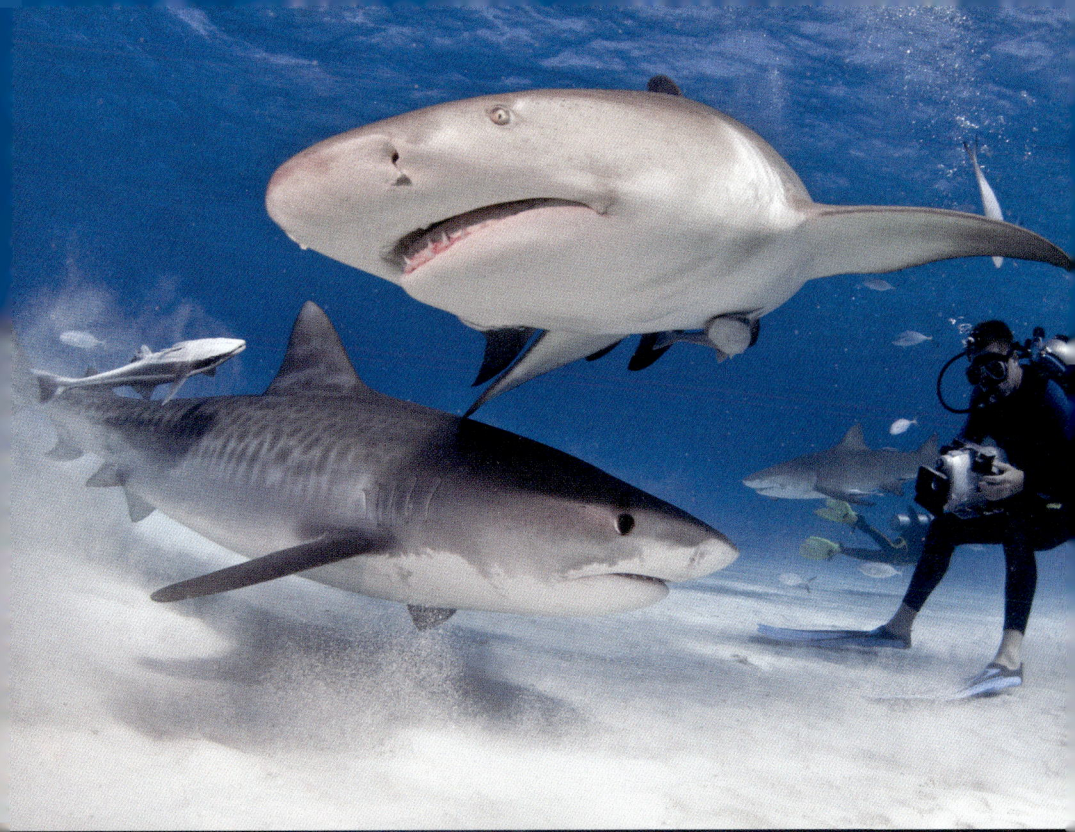

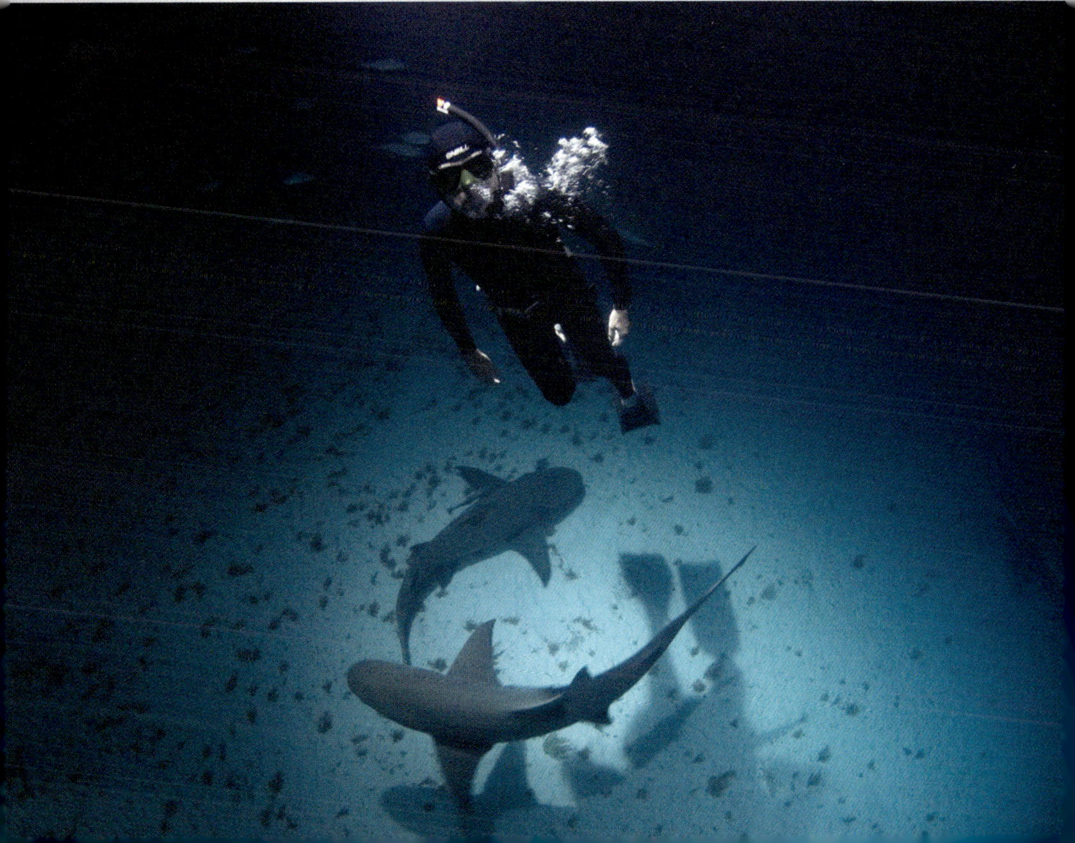

Clockwise from top left: Off the Bassas da India Atoll in the Mozambique Channel. • Diving with galapagos sharks in the Bassas da India lagoon: there were a lot of sharks about and we used protection. • Tiger sharks are among the most beautiful sharks in the ocean. • Blacktip sharks in southern African semi-tropical waters.

Above: Diving with lemon sharks in the Bahamas. **Below:** Aquariums might be controversial to some, but they have their educational role.

Above: Free-diving with bull sharks off Mozambique. • Lemon sharks in the Bahamas.
Below: Sharks in the evening light, just under the boat.

Above: Lemon shark. **Below:** A bull shark comes face to face with a free-diver.

The shark made several lunges at the divers, but somehow their spirited response kept it from biting them. Each time it came in they yelled and punched it with all their might. After what seemed like an eternity, but in reality was only six or seven minutes, they swam directly at the shark and it turned tail and vanished, leaving behind two relieved divers who nursed some badly bruised knuckles.

The first great white ever killed in South Africa by spearfishermen was a three-metre, 136 kg specimen subdued by the late Tony Dicks off Bird Island, in Algoa Bay. Dicks – he died not long afterwards in a tragic shallow-water blackout incident – speared the shark in the spine. Instead of disabling it, the wound enraged the shark and it attacked him.

Morné Hardenberg

Dicks escaped injury only by jamming his CO^2 speargun down its gullet and discharging a burst of gas into its body before swinging onto the shark's back to avoid its slashing teeth. Dive buddy Clive Tutton put another spear through the shark's gills to finish it off.

Historically, it has not only been divers that have had chilling encounters with the Submarine and other large sharks in False Bay.

For some reason, great whites that cruise the bay seem to have an intense dislike of boats. Indeed, it is so bad that the area has seen more attacks on fishing and ski-boats – shades of *Jaws* – than Australia and the United States combined. More than one boat has had chunks chomped out of it by an aggressive 'johnnie', which has even bitten and bent props and outboard motors.

Like their energetic cousins, mako sharks, great whites have been known to jump into boats while pursuing fish being reeled in by anglers. One vivid example has gone down in local angling folklore:

In January 1982, Theo Ferreira, then a dedicated shark hunter, hooked a four-metre great white early one morning while fishing off Macassar beach. Just when the exhausted shark was finally alongside the boat and ready to be roped, there was a surge of water and the huge head of the Submarine reared up and, with its cavernous jaws that were more than a metre wide, clamped on Ferreira's catch. He watched in stupefaction as the two-ton monster turned to swim away with a one-ton shark in its mouth.

When it felt Ferreira's restraining fishing line still attached to its prey, the Submarine leapt high out of the water, still with the other shark in its jaws, and then came crashing down in a welter of blood and foam to snap the trace.

'I was awed,' remembers Ferreira, dating his change from great white shark hunter to conservationist from the day of that unforgettable encounter.

While the American moviemaker Peter Gimbel was making his acclaimed 1971 film *Blue Water White Death*, he and his crew arrived in Cape Town to begin their cinematic seven-month voyage in search of the stars of the movie – great white death sharks, as they referred to them. Once in southern African waters, they fitted out their old converted whale chaser *Terrier III* for their worldwide shark hunt with Australia's legendary diving couple Ron and Valerie Taylor.

Gimbel told me at the time that it was actually not necessity for him and the others to embark on a round-the-world odyssey. His team had found all the great whites they needed to film in False Bay, right on Cape Town's doorstep. But the movie called for an unfolding global search to maintain the thread of drama.

One result of Gimbel's riveting documentary was to forever change the name of *Carcharodon carcharias* among local divers from the innocuous-sounding blue pointer to the more chilling great white. The subsequent Hollywood movie *Jaws* and its sequels helped to cement the name, as much as it spread the myth of the shark as a mindless man-eater.

While there is no doubt that the appearance of such large predators tends to instil fear among those who encounter these magnificent creatures in their natural element, scientific opinion seems to have come round to the view that, to quote my old friend Jeremy Cliff, all sharks are more sinned against than sinning. That is especially true of great whites and their cousins, collectively known as requiem sharks – from the first word of the Latin mass for departed souls.

Unlike other fish and in common with the thresher (*Alopias vulpinus*) and makos (*Isurus oxyrinchus*), great whites have a somewhat specialised circulatory system that effectively makes them partially warm-blooded, a biological bonus that gives them the edge over their cold-blooded brethren.

Though their 'warm' blood is only six or seven degrees Celsius above ambient water temperature, it makes for greater powers, makes for more efficient swimming muscles, a more hi-tech brain and equips great whites with better colour vision than the rest of the world's shark species.

It is also the only shark known to 'spy-hop' like a whale, sticking its head out of the water to see what's going on around it – as some of the Springbok spearos mentioned earlier in their False Bay encounter.

Great whites do not usually attack humans on sight, but if for some reason that happens, attacks in most cases are thought to be investigative, which is cold comfort if you are the subject of interest. Almost without exception, victims have been swimming or floating on the surface. These critters also constantly check out the surface while patrolling the depths at about five or six knots.

Additionally, they have outstanding eyesight. Even from 20 metres down they can spot surface objects as small as 15 centimetres across. And when visibility is down to a metre or less, you can be sure that if there is a great white in the vicinity, it is following your every move even if it cannot actually see you.

Its first bite – which is now referred to in the industry as 'mouthing' – informs the shark through taste and touch receptors whether the object is edible and worth a follow-up strike. It invariably lets go of anything with teeth or claws that might injure it in a struggle. This is why it has the disturbing habit of rolling its eyes back in their sockets and covering them with a protective lower eyelid known as a nictitating membrane.

Similarly, it does not take to the stuff of which wetsuits are manufactured, which is why so many divers – Tommy Botha included – have been 'mouthed' and immediately released.

One lasting impression remains fixed in the minds of so many divers who have tangled with these critters. When coupled with a formidable mouth gaping in what seems to be a huge grin, this sinister wink has given divers who have been close enough to witness it, the impression that the great white is actually quite pleased to see them.

When the shark does take a serious bite, its jaws can exert a force of up to 3,000 kg/cm^2, which helps explain why even the smallest bite is serious enough to be life-threatening. Nearly half of all shark attack victims die, usually from shock and blood loss. A simple 'mouthing' endured by Tommy Botha resulted in more than 100 stitches.

Divers have learnt to recognise the signs that indicate that a shark attack might be imminent. If the creature swims slowly around you it is probably only curious. The first sign of danger occurs when it begins to swim erratically. Then, if it hunches its back, shakes its head and drops its pectoral fins, you're in pretty serious trouble.

The exposed jaw of a great white flashes a frightening set of teeth. And to observe this from close quarters, the shark is right alongside and is definitely on the verge of attack. That's when you need a cool head and, if possible, a shark-billy or a Powerhead charge to fend off the attack. Screaming has been effective in keeping some of the less determined predators at bay, but by all accounts, it doesn't always work …

Authenticated shark attacks have been systematically recorded in southern Africa for more than half a century, though there are sketchy reports going back at least to 1886.

Two of the largest great whites ever seen in southern waters entered False Bay in the late 1800s, in the wake of a sailing ship which attracted these unwanted consorts by a constant succession of sea burials after an outbreak of plague on board. One of the monsters was eventually harpooned and killed by a volley of rifle fire. The carcass was hauled ashore at the bayside village of Glencairn and according to reports of the time, measured more than thirteen metres.

In the same era a great white of more than eleven metres was taken in Australian waters off Victoria, although it must be said that these historic records are viewed with some scepticism by marine biologists, who regard with suspicion claims of any specimen of over eight metres. Still, there must be some truth in some of the reports because they are pretty consistent.

A great white harpooned southeast of Durban by the whale catcher *Kos 32* in June 1962 was the largest ever recorded in southern African waters. This giant weighed three and a half tons and measured more than six metres. The largest ever landed by a shore-based angler, incidentally, was a great white beached below Durban's South Pier in 1953 by Reg Harrison. He landed the 754 kg shark on a line with a breaking strain of only 39 kg after an epic six-hour battle.

The largest ever great white netted was brought in by a Gansbaai trawler early in 1987 and tipped the scales at 1,214 kg. It was caught in the area where the famous British troopship HMS *Birkenhead* struck a rock off Danger Point in 1852 and foundered with the loss of 445 of the 638 soldiers and civilians on board. Many of the victims in that catastrophe were believed to have been taken by great whites and hammerheads which are still found around there.

Latter day heavyweights dwindle into insignificance when compared to their ancestors, mammoth great whites known to science as *Carcharodon megalodon*, a shark that weighed up to 20 tons and grew to more than 30 metres long. These were cruising the primeval seas in the Miocene period, more than 25 million years ago. Their fearsome fossilised mega-teeth have been dredged up from the depths and found in the fossil beds of KwaZulu-Natal and in the Western Cape, all wickedly serrated triangular and more than 15 centimetres long with weights of about a third of a kilo.

Such a shark would have had jaws of at least two metres wide and would have been capable of swallowing whole something as big as an ox.

They must have hunted prey as large as a baleen whale.

As mentioned in an earlier chapter, the year-end summer vacation in 1957 saw up-country holidaymakers pouring into Natal in their usual droves to enjoy the endless sea, sand and sun. It was a time forever to be remembered as 'Black Christmas', after an unprecedented series of shark attacks spread fear and led to thousands of vacationers abandoning coastal resort towns and villages in scenes that clearly foreshadowed those in the movie *Jaws* decades later. Peter Benchely had obviously done his homework.

The first attacks to spark the panic occurred in murky water, which is well known to attract sharks, zambies especially. A week before Christmas, 16-year-old Robert Wherley was swimming only 50 metres off the beach at Karridene when a shark struck. Doctors saved his life but he lost a leg.

At Uvongo, near Margate, 15-year-old Alan Green was so badly mauled by a shark that he died before he could be brought ashore.

The next fatality also happened near Margate three days later. Vernon Barry (23) was splashing in water barely a metre deep when his left hand was bitten clean off. His left leg and right arm were also savaged by the shark and he died in hospital soon after.

The fourth attack in twelve days also occurred at Margate. Pretty 14-year-old Julia Painting, on holiday from Bulawayo in Rhodesia, was frolicking in the surfline when a shark grabbed her. In spite of heroic rescue efforts by another Rhodesian holidaymaker, Paul Brokensha, the teenager was maimed for life in the attack. The shark tore off her left arm at the shoulder before Brokensha beat it off with his fists.

Ten days later, just when people were thinking the worst was over, the unthinkable happened. Derryck Prinsloo (42) was killed by a shark in turbid water only a metre deep and while he was barely five strides from the beach. During the Easter holidays three months later, attacks at Port Edward – a few kilometres south of Margate – led the authorities to respond by calling in the South African Navy frigate SAS *Vrystaat* to depth-charge the sea around Margate. Hundreds of dead fish floated to the surface after the underwater explosions, but no dead sharks were seen to surface.

The danger of a repetition of the attacks of 'Black Christmas' and the devastating effect they had on the local tourist industry, led to the creation in 1964 of the Natal Anti-Shark Measures Board – now the KwaZulu-Natal Sharks Board – which experimented with safety nets until nearly 50 popular beaches were protected by more than 40 kilometres of gill nets.

These have been extremely effective, netting hundreds of sharks a year since their introduction and so reducing the populations of large sharks off Natal.

Anti-shark netting was deemed necessary only on the Indian Ocean shores of South Africa and scenes of shark attack then became more likely further south and west. During the early years of World War II, sharks had already shown themselves to be no respecters of geographical or thermal boundaries.

Gary Haselau, a Cape Town diver who spent time as a photographer

Siimon de Waal

with Jacques-Yves Cousteau on his oceanographic research ship *Calypso*, recalls a November day in 1942 when he was walking along Clifton's popular Fourth Beach, watching two swimmers making for the shore. He saw 21-year-old medical student Willem Bergh in the waves close to the beach where hundreds of people were relaxing and sunbathing. Within 20 metres of the shore the young student was attacked by a shark that Haselau estimated to be possibly six or seven metres in length, undoubtedly a great white.

'Both his legs were severed with one bite,' says Haselau, 'and we all watched horrified as the shark slowly cruised out to sea with the head and arms of the dead student clearly visible above the waves in its jaws.'

In the years to follow, Haselau himself had a few serious encounters with a variety of sharks, but counts himself lucky he has never faced a shark like the one he saw that day at Clifton.

Geoffrey Spence was swimming off the same beach in November 1976 in water that was unusually warm and clear for the normally chilly Atlantic. Visibility was around 30 metres, but even so he never saw the shark racing in to attack him.

'I suddenly felt a hard thump on my side and then my chest was clamped in what felt like a vice,' says Spence, who was propelled through the water by the shark before it released him. It then slowly circled him within touching distance before swimming away.

Spence's memento of his frightening encounter was a pattern of punctures from the bottom of his ribs up to his shoulder blade.

The relatively shallow depth of the teeth marks indicated to an expert – who later examined him – that the shark had only 'mouthed' Spence in an investigatory attack. That victim was inordinately lucky.

The quiet little Western Cape village of Gansbaai ('Goose Bay') is known as the shark-watching capital of South Africa, where tourists pay for the ultimate thrill of seeing a great white face to face underwater, but from the safety of a stainless-steel cage.

Not far away lies Dyer Island, named after a Nantucket black man by the name of Sampson Dyer, who, as we have also seen in an earlier chapter, worked on the coffin-shaped island during the heyday of the American sealers in the early 19th century. Between Dyer and nearby Geyser Island is a narrow channel notorious as 'Shark Alley'. Many divers have tales of some grim encounters with sharks in this underwater thoroughfare.

Paulo Fossati

CHAPTER TWENTY

RAGGEDTOOTH SHARK: THE OLD LADY OF THE SEA

We have seen Geremy Cliff on the box often enough when he talks of his passion for sharks and the sea in some of the *Discovery Channel*'s programmes. Among his favourites is the raggedtooth shark (called sand tiger in America or grey nurse in Australia). Colloquially, we know it as the raggedtooth shark or 'raggie', because of its formidable rows of teeth. A few extracts from Geremy's notebook …

I took a deep slow breath while trying hard to relax to ensure a good bottom time. Colin had already speared a "cracker" of at least ten kilograms. We were both aware that there were even bigger fish lurking in the gullies.

My last shot had been aimed at one of the brutes hanging at the back of a shoal. Trouble was, my spear bounced harmlessly off one of its body scales behind the pectoral fin. It was a big critter and I was sorry to have missed.

I finned slowly downward, my speargun tightly clenched in my right hand. As was usually the case – whenever I went deep, the fingers on my left hand were on my nose-piece to equalise. I was very much aware that there were sharks about: we'd seen several and I couldn't help the surge of anticipation at the prospect of encountering my first 'raggie' of the day.

The reef below slowly took shape before me. A small shoal of zebra-fish darted over the edge of the drop-off into deep water. I levelled off, lowered my legs and lifted my head. As I looked up, I spotted a large shark swimming towards me at its characteristic relaxed pace.

There is nothing hurried about raggedtooth sharks, among the most beautiful and graceful creatures in the ocean, except, of course, when they're disturbed or in hunting mode. Its pointed snout

Sijmon de Waal

remained steady while its long tail fanned the water with deliberate, methodical beats.

I knew the score, having been there often enough before in Indian Ocean waters. If I kept myself totally immobile, the sand tiger would veer off to the side. They tend to do that just before they reach touching distance. This time, it seemed to keep on coming towards me, its approach fairly slow but, from up close, pretty determined. I wasn't comfortable with the encounter.

I was suddenly also aware there was absolutely no chance of my avoiding the shark if it didn't turn away. Time seemed to stand still, like it sometimes does when you're in a hurry and waiting for the lights to change.

By now it was too late to raise my gun. I jerked my head back in the hope that this sudden movement would cause the shark to change course, at the same time kicking out rather awkwardly with my fins. I needed my body to follow, but everything seemed to be moving at quarter-speed.

Finally, I kind of succeeded, only to feel a glancing blow on my right hip, coupled to a swirl immediately in front of my face that knocked my dive mask sideways. What had happened was that the shark had swum right into me and by now my mask was flooded. Additionally, I was grappling with more immediate matters like being able to see again: I had no idea where the shark was or what it was going to do next.

The quick answer might have been to get the hell out of there, but then when you're underwater and can't even properly make out the blotches on your hand, you simply don't just barge blindly about. I kicked out for the surface, my heart racing and gasping for air.

As I broke the surface I tore off my mask. Moments later I was greeted by my buddy, whose face was creased by a wry grin.

'Don't take chances with our "raggies," he remonstrated quietly. 'They're not the docile beasts you'll find along your coast … Natal is very different,' he added.

When I'd regained my composure, Colin explained that the shark had simply carried on along its original course after bumping me out of the way. It hardly took any notice of me as it moved on, he added.

This brief, but alarming experience – though it happened in southern African waters nearly 30 years ago – remains etched in my mind. Looking back, I am also aware that an important lesson had been imparted that sunny summer morning off Plettenberg Bay.

I was young and impetuous. I also believed that nothing was beyond my ability. In fact, I had become over-confident and blasé to many of the perils that awaited any over-complacent, cocksure young diver. Like my fellow aquanauts, I was an intruder in the marine environment and the sooner I developed a proper respect for the animals that eventually evolved into a lifestyle that was designed to cope with the world beneath the waves, the better.

I had to accept too, that once I'd donned my mask and flippers and plunged into the sea I could no longer regard myself as superior or even equal to the most docile of raggedtooth or sand tiger sharks.

Now, after a couple of decades as a shark biologist, my experiences have given me a good insight into the world of sharks and their domain. It is an awareness that has engendered both admiration and respect for this group of creatures normally scorned by man as the vermin of the sea. How absolutely wrong!

Having subjugated almost all the great predators that roam the surface of the earth, we are now obsessed with the desire to annihilate the maritime beast that we fear the most and over which we have little or no control – the shark.

The fact that there are over 400 species of shark, most of which are no more than a metre in length (and often as timid as your average lap dog) is often disregarded. Leonard Compagno, one of the world's most respected shark researchers, estimated that less than a quarter of the roughly one hundred species that occur off the South African coast offer any form of threat to man.

Victimised largely due to its deceptively ferocious appearance – in Australia, this once-common shark was decimated by 'sport divers' armed with explosive Powerheads – Australia declared it a federally protected species in 1997. As a result, these shark populations are slowly beginning to recover in some areas, but are still not nearly as abundant as they once were.

The sand tiger shark can be dangerous. But looking back on the little incident in Cape waters, I wasn't able to interpret that creature's behaviour as possible aggression. I was simply in the way of a shark that had no intention of deviating from its path: and I realise today that I was fortunate indeed not to have been gashed by an array of needle-like teeth that are the most significant feature of any of this shark's jaws.

Unlike most bony fishes, which have teeth firmly embedded in sockets within their jaws, the grey nurse shark's teeth are only held in connective tissue stretched over the cartilage of the jaw. A good set of teeth is as important to any predator as fringed sheets of baleen might be to a feeding humpback whale.

With the sand tiger – behind its impressive front row of jagged, dagger-like teeth – lie at least four more deadly rows that move steadily forward over time to replace those that are damaged or lost. Scuba divers searching the sandy gullies on reefs at Sodwana, Protea Banks and Aliwal Shoal often come across 'raggie' teeth: they're easily recognised by their awl shape with a small spine, or basal denticle on either side of the main cusp.

During the lifespan of such a shark, which may be several decades, over 1,000 teeth are produced and lost.

The sand tiger, or in our parlance the raggedtooth shark, is known to scientists as *Carcharias taurus*. It is widely distributed in both the northern and southern hemispheres. Called the grey nurse shark in Australia, it is one of the commonest of the larger predators in the inshore waters of southern Africa and may just as easily be found on shallow reefs between Plettenberg Bay in the south or off Inhaca Island in southern Mozambique.

Initial perceptions among divers are that this is a sluggish predator. But you need to exercise caution.

Its slow speed belies astonishing agility and acceleration when an unsuspecting fish ventures too close to those jaws that bristle with teeth. It can be equally uncompromising when cornered, as was the case when Al Venter recently ventured into one of the 'Raggie Caves' on Aliwal Shoal.

'I was caught completely off-guard,' recounted Venter afterwards. 'I'd gone a metre or two into the cave with a narrow entrance and a fairly low overhang. Once my eyes became accustomed to the gloom, I spotted a rather large raggedtooth shark at the far end. It didn't worry me unduly, because I could easily turn around and head out to open water.

'But before that could happen, the shark headed straight towards me and I was suddenly made aware of the rather extreme confines in which I was "trapped". I knew that I couldn't just turn and flee: that might have made this predator aggressive: after all this was a large creature of about three metres. Nor could I duck low so that the shark could get past over my head: I was already on the bottom …

'It could feasibly squeeze out beside me, but that would be the shark's decision, not mine and in any event, I wasn't sure there was enough space.

'The raggedtooth kept coming and once it was directly in front of me; it surprised me even more by turning and facing me head-on: I'd never had that happen before! I tried backing out, but that would have needed quite a bit of arm action in front of a shark of uncertain predisposition, so I wasn't very successful there either.

'For a few moments we stared at each other, the shark's large head half-an-arm's-length from mine. Jeez!

'Then, almost as if it had been rehearsed, there was a loud clap – which is symptomatic of 'raggies' when they take fright – and the next moment the shark had forced its way past me and was headed out of the cave and into open waters.

'I was mightily pleased and not a little disorientated …'

The diet of the raggedtooth shark is composed almost entirely of fish, supplemented by the occasional small shark and squid. These include inshore dwellers such as mullet and springer and englishmen, slinger and other inhabitants of deep-water reefs. Likewise with small tuna and bonito, which are able to dart with remarkable agility but are often still impaled on the sand tiger's teeth before being swallowed whole.

These sharks often feed collectively, a phenomenon that we've witnessed among the more highly evolved dolphins.

In Australia a diver once observed a pack of these sharks herding a small shoal of unhappy kingfish for breakfast. They did this by thumping their tails to generate pressure waves and which, to the initiated, sounded almost like gunshots.

While most large sharks are committed to a life of perpetual motion (to ensure that oxygenated water passes over their gills) there are a few that do not. The sand tiger actively pumps water over its gills, which enables it to reduce the energy needed to maintain a high speed and which is why they are often found 'sleeping' in caves.

But this, in itself, gives rise to another problem: sharks are negatively buoyant and lack the swim-bladder found in most bony fish, which helps to maintain neutral buoyancy at any depth. The oil reservoirs stored in its enormous liver do help, but they are insufficient to prevent the shark from sinking to the bottom as forward motion is lost.

In 1972 John Bass and John Ballard of the Oceanographic Institute in Durban discovered the sand tiger's neat answer to this problem. They observed sharks swimming to the surface in the shark tank, where they swallowed gulps of air. This air trapped in the stomach effectively acted as a swim-bladder and provided positive buoyancy.

An interesting aspect to the life cycle of this predator is the migration of mature sand tiger females each spring into KwaZulu-Natal waters in order to mate. By then they would have travelled all

Sean Botha

the way up the coast from the colder waters of the Cape to meet their male counterparts, whose whereabouts before and after conjoining remains a mystery.

Mating, I was discovering for myself not long afterwards, is certainly not gentle or affectionate. Females, including the occasional adolescent, are covered with courtship bites inflicted by the males to determine whether a female is sexually receptive. And while the act of copulation is rarely observed in large sharks, it is thought that the male 'raggie' seizes the pectoral fin of its partner with his teeth as they swim slowly in mid-water, with consequent close contact between the genital regions.

The male then inserts one of his claspers (the paired intromittent organs between the pelvic fins) into the urogenital opening of the female and the transfer of sperm takes place. The presence of a pair of these claspers on all male sharks is the certain indication of their gender.

After mating, the females continue their migration northward, usually remaining quite close inshore. By midsummer they can be found in packs on the shallow reefs of the clear waters of northern Natal where they spend much of their gestation: presumably the warm waters stimulate the development of their embryos.

Many divers originally believed that the females congregate in these places to give birth, but research has shown that the nursery grounds are confined to cooler waters, and the females need to make the long journey south to the Eastern Cape before they can drop their young in midwinter.

Further amazing discoveries have been made about the mode of reproduction of the sand tiger shark.

In 1948, the American biologist Stewart Springer found a number of pups of varying sizes in each of the two uteri. This in itself was not unusual, for many sharks are viviparous, producing between six and twenty offspring.

It was only when he cut open the largest pups that he discovered that their stomachs were full of egg yolk. It became apparent that the mother produces masses of eggs throughout a pregnancy of something like nine to ten months.

A diligent researcher once counted over 24,000 eggs in the ovary of a pregnant sand tiger that was almost three metres long. Each egg, roughly the size of a pea, is packed along with five to ten others in a small bag. A developing embryo may emerge from each of the earliest egg parcels, while later packages contain only unfertilised eggs. These, in turn, create a nutritional diet for the developing pups, which gorge themselves with egg yolk.

But all that still did not explain why the 'raggie' litters habitually consist of only two pups (one in each uterus).

Springer discovered that the embryo that develops first in the litter has a head start over the others. Its rate of growth is accelerated by ingestion of the egg parcels to such an extent that it soon becomes large enough to devour its siblings, until it remains the sole embryo in the uterus.

This bizarre event is typical of the survival of the fittest throughout the animal kingdom. The result is that by the time the two young sand tigers are born in the nursery grounds of the Eastern Cape, each about a metre long and weighing about eight kilograms, they are already efficient hunters. With their keen sense of smell and a well-developed set of teeth the newborn sharks can easily fend for themselves.

Sharks, like most of the other lower vertebrates, display no parental care, and their offspring must be able to avoid predation from birth. Stores of fat in the young sand tiger's liver and the remnants of the last meal of eggs swallowed (while it was still inside the mother's uterus) will keep it going until it can catch its first prey.

Female sand tigers are not cannibalistic, but many other female sharks stop feeding when they enter the nursery grounds to give birth. These mothers then leave immediately after pupping to avoid feeding on their own offspring. It is important for the female to replenish the depleted energy reserves in her liver and muscles after giving birth, for the demands of nourishing offspring for nearly a year are great. Recent research has shown that the females do not reproduce every year; presumably an interval of a year between pregnancies enables the sharks to build up their store of energy.

A mystery that remains unsolved is the lifespan of the grey nurse. Specimens may attain a length of about three metres and the heaviest South African individual is about 300 kilograms – undoubtedly female – as they grow larger than the males in many species.

Ageing studies of sharks have proved difficult. Scientists count the concentric growth rings of the disc-shaped vertebrae (much as the age of a tree can be estimated by counting the rings in the trunk). Kenneth Goldman and colleagues determined age at maturity for northern Atlantic sand tiger sharks at being six to seven years for males and nine to ten years for females. Consequently, this is regarded as one of the faster growing sharks. What is of interest though is the longevity or lifespan of this species.

Sand tiger sharks have been kept in captivity for as long as 20 years. During the course of my own research efforts – that have involved regular tagging of pregnant females at Sodwana Bay and at Leven Point, 40 kilometres further south in the 1980s and 1990s – is that we are now witnessing a recapture of those same females, more than 20 years after they had first been tagged. This indicates that sand tigers have a lifespan of more than 30 years.

Despite their size, these large creatures tend to thrive in aquariums. Their slow speed soon enables them to adapt themselves to the confines of a tank, and since their diet is primarily fish, they are easy to feed.

Sharks have always been regarded as primitive animals, with a narrow repertoire of behaviour imposed by a small brain. That is not necessarily so.

An animal behaviourist, Anne Alexander, while working at the Oceanographic Research Institute in Durban, was able to train a 'raggie' to surface and take a morsel of food from her hand.

Several sand tigers have been born in captivity, a fascinating experience for those privileged to observe it. Unlike most other shark species, these pups are born head first, but only after they have turned 180 degrees inside the mother, for they spend the last few months of their development facing forward.

Others have been born by Caesarean section, removed from their mothers, who have died after capture. In such cases the pups may need some help, depending on how premature the birth is. Once removed from the mother, the pair of newly emerged shark pups are completely limp and unable to swim. Up to two hours may be needed to get them mobile, during which time the swimming muscles will tone up, the fins stiffen and the pups get their bearings in their brand-new environment.

As we're all aware by now, encounters between raggedtooth sharks and spearfishermen are not uncommon in southern hemisphere waters. In most cases divers aren't troubled until they begin to shoot their fish. Not surprisingly, blood in the water – coupled to the struggles and gyrations of dying fish – tend to arouse a lot of interest in predators. Many, particularly hammerheads and small dusky sharks, will quickly attack speared fish.

The 'raggie', in contrast, will often ignore the prospect of an easy meal dangling from the end of a spear or fish stringer and show aggression to the diver himself. Territorial aggression and resentment of another predator are possible explanations for such behaviour. Fortunately such incidents rarely result in injury.

Sensible divers are likely to leave the area and respect the prior rights of the shark when this happens. Others who wish to show off or believe that the only good shark is a dead shark, may use an armed Bangstick with a rifle bullet or shot-gun cartridge in its breach to kill or maim the creature.

Such activities were apparently in vogue in Australia in the 1950s. Ben Cropp, a spearfisherman who was responsible for the development of the Powerhead, together with his companions ended up killing 18 of these predators, most of them sand tiger sharks from the same stretch of reef in two days. All died instantly.

In his book *Shark Hunters*, Cropp concluded that '… the grey nurse is being methodically wiped out in New South Wales, and will soon be a rare shark in those waters.'

In recent years, I have had many encounters with large female grey nurse sharks in the clear warm waters of Maputaland. Each time I've been struck by the serene air of these animals as they glide slowly and effortlessly over the reef.

I have often wondered how these sharks regard me as they peer out of their suspicious, unblinking eyes. Often I have come to realise that my presence has disturbed one of them, as with a single swish of a tail the streamlined form darts away. Others will merely turn away slowly before they reach 'touching' distance.

For my part, I have to resist the temptation to violate the golden rule and not reach out and stroke the shark as it turns away …

The raggedtooth shark, like all predators, deserves nothing but admiration as one of the most efficient predators of the sea.

UPDATE: Listed as 'vulnerable' (VU A1ab+2d) in IUCN Red Data List, the sand tiger was declared a protected species in the western North Atlantic by US National Marine Fisheries Service (NMFS 50 CFR, Pt. 678), April 1997.

The status of the sand tiger shark in Australia (where it is known as the grey nurse shark) has been reclassified. There are now only about 500 individuals left on the east coast and are listed as endangered or critically endangered depending on whether you're looking at State or Federal legislation.

CHAPTER TWENTY ONE

CAPE TOWN'S TOMMY BOTHA AND HIS SHARKS

Tommy Botha has had more encounters with great white sharks than most other divers in the country, if not the world. While spearfishing, he has been harassed by them dozens of times. Once, off Cape Point, he was physically thrust – at speed – through the water by a four- or five-metre great white that came up from below like the proverbial bullet. It apparently did not like the taste of his neoprene wetsuit so it did not go in for the kill.

That attack from below was in the classic feeding mode of the great white shark when hunting for seals. Clearly, it initially believed Botha to be a seal.

Since then, he has been attacked again. The last time he needed 30 stitches in his hand after a great white shark towed him through the water for about a minute, obviously puzzled by the behaviour of its victim in not panicking violently as seals are wont to do.

Tommy Botha estimates that he must have seen hundreds of great white sharks in the water while spearfishing over much of past decades.

The author, Al J. Venter – who has dived salvage at De Dam on the Cape south coast on the wreck of the British East Indian *Johanna* (1682) with Tommy Botha and his buddies Gavin Clackworthy and Bert Kutzer – interviewed this intrepid undersea adventurer.

AJV: Why do you think that you weren't physically mauled like 21-year-old Monique Price in Mossel Bay in June 1990?

TB: Well I was bitten on the hand which is really not as bad as a leg attack. The shark took my hand in its mouth and swam in a circle.

Sean Botha

With your hand in its mouth?

Yes. When I hit it on the side of the head, it let go.

Why didn't it bite off your hand? It can easily do that.

My hand was actually in the corner of the mouth, so it wasn't a direct bite. Also, teeth on the sides of its jaws are a lot smaller than in the front.

Did you need stitches?

A lot, but it has healed completely.

How big was it?

Not that big as whites go, I reckoned it to be about three metres at the time, possibly four. I was actually quite lucky because it could have been much bigger. Some great whites in our waters are easily five metres and more than a metre across.

What is your theory about the increased numbers of great white sharks in Cape waters? There are a good many more in False Bay than there were before.

The answer seems to lie with the reality that they have no enemies. Put it down to their fearing nothing, though that said, all sharks are inherently cautious. Ten or 15 years ago, it was unusual for any diver to see a great white while in the water. If you did spot one, you talked about it afterwards. But these days, if you are an active diver, everybody has either seen one or more of them. At Struisbaai you can see a couple of different ones on a good day's diving and great whites are definitely becoming more plentiful.

But surely they have been here all the time. What were their enemies 15 years ago?

I don't think there were that many 15 years ago. They have just grown bigger and, obviously, they've multiplied. And they haven't been culled because it is against the law and there are no Chinese fishing boats around. Also, it seems as if fish have been getting fewer and fewer. They live off fish. And seals, of course.

While divers are seeing more great whites just about every weekend, why haven't divers had more bad experiences with great whites?

If you actually see a great white in the water with you, you can consider yourself fairly safe. It's when you don't see them that is the problem. Just about everybody that has been attacked never had an inkling that there were sharks around before they were hit, almost always from below. But if they swim round you – and you behave differently to the creatures they usually feed on – then you still have the option to leave the water. Also, if you're alert and keep your speargun between you and the shark, you are relatively safe. And then you need to get into the boat. Quickly!

But the attack on Monique Price, as far as we know, was the first attack on a scuba diver in southern African waters, even though scuba divers have been attacked by great whites in Australia.

The reason could be because scuba divers are normally under the surface of the sea. Monique was attacked because she was still alongside the boat and on the surface. If you are underwater it is different: you are within the shark's vision. That's not the case when you thrash about on the surface, which might be why she got mauled. I haven't studied the details of Monique's tragedy.

Sijmon de Waal

Jerry Buirski, who dives a lot when he's not out at sea, tells me there are a lot more great white sharks in the whole False Bay area, including such places as the Rocky Banks and Whittle Rock.

I agree. I think there are more at Whittle Rock and there are definitely many more great whites inside the bay than on the outside, which is almost certainly because of the seal population of the bay. We don't see many at Anvil or at Bellows Rock, closer to the tip of the Peninsula.

252

Do you think that the great white is territorial?

Yes, I firmly believe it is. You always seem to find them at certain places and there are those who say that you get the same great white at the same place weekend after weekend. This is something borne out in part by the kind of research now being done on whites by Alison Kock.

Whittle Rock used to have a 'resident' great white some years ago. Is it still around?

Divers were always being buzzed by it in the past, but I haven't been there for a long time. There are those who claim it is still around, or something large in any event. We never dived there unless it was clean.

How do the great white sharks normally react if they are uneasy or aggressive towards divers?

They usually swim fast and then you really have to be on the alert. When a great white swims straight at you, or flits past you and turns in a matter of seconds before it is facing you again, then it's time to get back into the boat.

It is said that one of the signals emitted by an agitated great white is a swishing tail.

Well, normally their tails are so big that if one comes past, you can't miss getting the message, or for that matter, the swish. I've experienced a swishing tail that swirled my legs out from underneath me and before I could look properly at the shark, it was facing me, jaws open and pushing me forcefully backwards through the water. It all happened so quickly ... these sharks are enormously powerful and can turn on a sixpence.

But that time it didn't bite you, and, I suppose, it could easily have done so?

Quite right, it was a big shark, but it just kept pushing me and my gun backwards through the water while I was trying to get away from it. That was a five-metre monster; it could have probably swallowed me whole. And then it just left us and swam slowly away. All three of us divers rushed for the boat!

You have rushed for the boat a few times in your life!

Quite a few times, but on that occasion I was the first on board.

Do you think we will see the return of the Powerhead as a means of diver protection?

Without question! The trouble with the Powerhead, strictly speaking, is that it is a firearm: you need a licence to own one. So you have to apply formally for that at a police station. Also, they're not cheap. Most divers that go out regularly these days take a Powerhead, or their own home-made versions with them. These are easily fitted to the spearhead if a shark threatens to attack or take your fish from you.

What calibre is preferred?

Some divers prefer the .357 Magnum handgun cartridge, but there are some that like 9mm Parabellum. I reckon that the larger calibres are more effective.

How effective are they against big sharks?

There have been serious problems with larger bronze whalers in the Struisbaai area. Some sharks have been taking spearo catches and have been in the three to four metre range. Once a fish has been shot, these bronzies come right up to us and try to steal the fish out of our hands and they quite often do so before we are able to get back to the boat. Some divers think these are great whites but they are not, though there are enough great whites in the Struisbaai area.

Are they aggressive?

Very! We've had to kill quite a few of these big sharks to try and get rid of them, or at least deplete their numbers. They are aggressive and some of the bigger animals fear nothing, which means that we have no option but to kill them. But what makes the bronze whaler very different from a white is that it will swim below you, giving you an opportunity to hit it on the head with a Powerhead: you aim for the area just behind its eyes, where its brain is. Effectively, you are knocking out its central nervous system and it dies very quickly. The great white rarely swims just below you ... it is usually right alongside you if it is at all aggressive or inquisitive.

Why next to you?

There is no particular reason, except that some of them quickly come right up to your level. And they will remain alongside you if they're inquisitive, which many of them are. That's why it is so difficult to kill them with a Powerhead; we can rarely get just above them to deliver a head shot, as is the case with a bronze whaler or a raggedtooth.

What do the divers do to counter this trend?

At Struisbaai when we are hunting yellowtail, we divers swim very close together – all of us on the surface with one of us diving down on the hunt. The others watch over him. Normally, when you get to a shoal of game fish, everyone will have a go at trying to shoot fish. But then, with a shoal in the area, there are also big bronze whalers and, more often than not, great whites around. Then one or two of us will have put Powerheads onto our guns while the others shoot fish. In this way we can sometimes react against sharks and not lose our fish.

What effect does the Powerhead have?

It kills a bronze whaler outright. I have hit a big one of up to four metres and it died immediately.

What was the reaction?

Once hit, the shark just dives straight down. The bottom is normally dirty and you don't know what happens to them. But they certainly stop hassling us.

What effect does this have on other sharks in the area?

We rarely see them in packs – usually on their own or perhaps in pairs. It is always when you shoot a fish on the bottom and start to pull it up that a shark seems to come at you out of the darker, colder water at the bottom. We don't usually see them from the surface until they are heading in our direction.

Tommy Botha, who lives across the bay from Cape Town, has set the scene for three generations of South African divers. That includes his children – both of them accomplished spearfishermen.

It is not generally known that this intrepid adventurer has worked salvage on shipwrecks in most of the world's oceans. He has raised treasure from Spanish galleons off the shores of the Cape Verde Islands as well as off Cuba in the Caribbean and still more in the Indian Ocean. Added to that, the Botha legacy is linked to scores of South African shipwrecks.

The man has never sought publicity, which means that he doesn't like to talk about sharks and shark attacks, in part because he has survived several attacks himself. However, after a few ales and a bit of coercion, he disclosed that apart from having been attacked and bitten, there have been many other times when he has simply avoided any kind of confrontation by getting back on board the boat.

As he says, he has been bumped often enough by whites. He suspects that they were 'mouthing' him, as the specialists like to call it – the shark 'tasting' its prospective target to establish whether it is edible or not.

'You have exactly the same thing when you see a great white come up from below and hurl a seal a metre or two out of the water. Only when the seal is on its way down again does the shark kill it, usually biting the creature in half or tearing off a chunk of flesh in its jaws.'

In Tommy's case – and with other spearos as well, these 'mouthings' were commonplace. In fact, he reckons, they still are.

'If this weren't so,' he says, 'there would have been many more divers killed by sharks.'

When Tommy's hand was grabbed by a five-metre great white shark, he hadn't seen his attacker beforehand. This is the way these things usually happen. Yet, he maintains, if you manage to make eye contact, chances are that the shark will veer away, though there is no guarantee that this will happen either. That fact is attested by several spearos who survived shark attacks and the few that fought for their lives and didn't come out of it alive.

'With my own attack, I was concentrating on a fish when the shark was suddenly there and grabbed me. It had my right hand firmly in its jaws and it dragged me through the water, and at quite at a pace too. Obviously, I tried to pull myself free, but it wouldn't let go. More to the point, why should it have? I was its next meal …

'Then I tried sticking my thumb into one of its eye sockets and beating it with my free hand.

'Fortunately, the glove I was wearing was made of some thick, resilient material and that helped. But it didn't prevent the lacerations that needed stitches.'

CHAPTER TWENTY TWO

GRACEFUL, THREATENED DENIZEN: THE GREAT WHITE SHARK

Gletwyn Rubidge of Port Elizabeth recounts his experiences with great white sharks while spearfishing one quiet Saturday afternoon. With a small group of diving enthusiasts, he went out to RIY Banks and dived along its western fringe. It was somewhat cold, visibility was moderate and there were no decent fish in the water. He tried chumming to attract fish, but still had no luck.

It was a fairly uneventful day and though there were plenty of fish about, nothing excited me. After about an hour, having decided it was time to move, I took a lungful of air and headed down in a final plunge.

At about ten metres, a movement to my left attracted my attention. In a fairly wide gully at about 20 metres I saw a rather wide profile heading in my direction. The 'visitor' was a great white shark, the operative word being 'great' because it must have been at least five metres long, or, in more realistic terms, more than a ton of shark! Instinctively, I pointed my gun at it.

What was disconcerting just then, was that this enormous shark was headed straight at me, all the while, with each stroke of its tail, shaking its head. I know enough about these predators to realise that this kind of head action – coupled with sharply downward pectoral fins – is a certain indication that the shark is about to attack. Clearly, I was in trouble, because as it got closer, it gathered speed.

Now I was really in the dwang! My thoughts just then were that I'd have to play my cards just right ...

I continued to maintain eye contact with the creature, which was kind of strange because of the way such an approaching shark looks when viewed from up-front; almost like an oversized ice-cream cone pasted onto a beach ball. Even at that slanted angle, I could clearly see both its eyes. And though all this activity had taken place within a matter of perhaps five or six seconds, it suddenly seemed as if everything had gone into slow motion. I never even had time to attach my Powerhead onto the tip of my spear.

It was very close by the time I decided that the best option would be to move forward towards the beast, as in a full-frontal 'confrontation'.

That was when I shoved hard with my speargun and pushed it firmly onto the head of the oncoming shark, whose jaws by now were wide open. The spear tip struck true and hard, right between the shark's nostril and right eye. And though I wasn't to know it yet, that seven-millimetre steel rod bent into a zigzag shape as I was forced away by the shark's momentum. With that, the monster moved away and out of range of my vision.

Surfacing quickly, I stuck my head out of the water to warn the other four divers who had originally entered the water with me. That done, I shoved my head back into the water again in case it returned. Slowly, very deliberately, I headed for the boat which was some 15 metres away. Having boarded, I again shouted at my dive buddies to get themselves out of the water. Everybody responded, except for Eugene.

It was then that we became aware that the great white was making passes at Eugene, even though he was trying to get to the boat: it was actually following him. He managed to board in the end, and then it was big smiles all round.

It is not every day that we meet such big 'man-eaters', even in the usually-friendly waters off Port Elizabeth.

For the rest of the outing, we decided to move to the far side of the RIY Banks. By now, like the others in the team, I was a bit nervous: all of us couldn't help looking over our shoulders each time we shot a fish.

Looking back afterwards, I knew that I was fortunate to have spotted the shark in its approach and to have had my speargun handy. Had it approached unseen, and from behind, it probably would have bumped me hard before taking an exploratory bite (as is shark custom when 'mouthing' potential targets). For its size, that hit would almost certainly have been a rib-breaker.

Cumulatively, over the years, I'd spent perhaps 30 days diving on RIY Banks: in that time I have seen hoards of bronze whalers, duskies and raggedtooth sharks, but this was the first great white that buzzed me in more than a dozen years of spearfishing.

Meantime, there have been quite a few other divers who have had close encounters with great whites. Basie Ackermann tells how, while spearfishing, he and fellow divers encountered a great white whilst diving. Richard Rumble, my brother David, and I were diving on Aliwal Shoal in the winter of 1996. Pilot shoals of the annual sardine run had already passed, but there had been no further action at the time.

We decided to try the shoal further inshore, but the visibility was terrible. By about six that evening it had got worse and visibility had dropped quickly from about ten metres to five or six. More out of the eternal trait of spearo optimism than much else, Richard and I decided to make a final drift dive over the shallow pinnacles. I could just make out the bottom below at roughly six metres.

I had my 'flasher' out and waited for something sizeable to appear. After about ten minutes I had to accept that this was a waste of time: there were almost no fish around.

I was just about to call the boat, when I saw movement towards the edge of my visibility. I stared hard, trying to make out what it was, but all I could see, and with some difficulty, was a lighter and darker line about two metres long that was moving quite fast. For a moment I thought I might be looking at a sailfish or something: it had that kind of shape, but also not quite …

I saw a dark top and a whiter bottom, but still no real shape, which should have offered a clue. Wondering what the heck it was, I stared intently.

Then it clicked. Figuratively speaking, I was looking at this apparition with a zoom lens, instead of a wide-angle focus, and it was only when I looked from left to right and took in the whole four and a half metres of it, that I realised that what was before me was a monster great white. It was so big that it completely filled my limited underwater vision.

It was not only the biggest shark I'd encountered until then, it was also my very first great white. Moreover, I had never encountered one of these creatures before, so I slowly panned towards my right and to its rear and there it was – the unmistakable tail. What I'd been staring at so intently in that limited visibility, was the very distinct line along its side, above its white belly.

Obviously, the reality of being in the water with such a huge shark worried me. Also, at six metres' depth, I was in comparatively shallow water. All that, coupled with six metres of visibility and it meant that I was in serious trouble.

This marvellously graceful creature – enormous as it was – was by then very close to me and I couldn't help feeling that suddenly the ocean had become too small for both of us. What struck me too was its awesome girth, which was almost hippo-like. Somebody else likened it to a small old-time Volkswagen.

I was able to reflect on the experience afterwards and, in retrospect the shark just looked so unbelievably powerful and majestic. In contrast, I just remember feeling powerless, and that had it attacked, my speargun would probably have been of no use whatsoever.

While I was still in the water I just prayed that it would keep on swimming past, but that didn't happen either. It was clearly attracted by my 'flasher' still merrily emitting its beams all over the place, when it turned towards me, angling up slightly from below where I was swimming on the surface.

By now I was shouting for the boat through my snorkel because the beast was getting very close. With other sharks in similar situations, I would always dive down and confront them, but this creature, in a word, was awesome.

When it came to within about three metres of me, I heard the boat pull up alongside and then it was my turn to do something miraculous: In what must have been the most fluid and beautiful single relocation movement of my existence, I emerged from the sea almost without even touching sides.

Richard, who was about ten metres from the boat and never even saw the shark, was equally quick to get on board once he'd been made aware of the problem. Nothing motivates a diver to move with alacrity through the water and back to the boat more than the prospect of encountering one of those big 'uns.

Although the incident couldn't have lasted more than a minute, I was able to get a good look at the shark. While it wasn't the biggest great white in the ocean, it was pretty long and had an enormous girth. The beam of our boat was approximately a metre and a half and I reckoned that the body of the shark, between one pectoral fin and the other, was actually broader. I couldn't even begin to guess its girth.

What did impress me was that this magnificent brute was not at all aggressive. I'd like to think that it just popped by to say hello.

On another occasion John Little was the focal point of a run in with sharks. Kirk Spilsbury, Dave Packer, and I were shore-diving at Leven Point off the Zululand coast in the summer of 2000.

We entered the water roughly in line with the sanctuary beacon about an hour after sunrise and went our separate ways. I headed straight for the outside edge of the shallow reef in about six or eight metres of water where we'd spotted shoals of 'cuda the day before. The current was north-to-south, while visibility was perhaps ten or twelve metres.

I was drifting along with the current when I saw a 'cuda swimming along the bottom: it was heading in from the other direction. I quickly turned about and saw that Dave and Kirk were about 100 metres up-current, but on roughly the same drift as I was.

I'd been looking away from the reef across a fairly sandy bottom and as I turned back I suddenly saw something grey emerge out of the gloom, followed by some oversized gill slits that were right by my head. That I was taken by surprise is understating the obvious: this 'thing' was suddenly right next to me – and then it was the shark's turn to take fright.

The creature bolted, but then, quite unexpectedly, it must have thought the better of it and did an abrupt about-turn. The next moment it was back, right alongside me.

I recognised the shark as a great white that was probably in the four-metre range, though in those few pensive moments it seemed a lot bigger, but then that's how these situations sometimes are. Undeterred this time round, I quickly loaded a Powerhead, took aim at the top of its snout as it rapidly closed the distance between us and fired.

The unthinkable happened: the Powerhead was a dud and it didn't detonate. Instead the shark took off and swam straight out to sea. It was a pretty scary experience that left me a bit shaken while I waited for Kirk and Dave to catch up. That was when we decided to continue with our dive, but to buddy-up and remain together.

Interestingly, there were two other divers, Ryan van der Merwe and Dan Owens, who were also diving from the shore off Leven Point that day. After we'd emerged from the water, they did another drift after we'd told them of the white. Both spearos got some fish, but Dan was pulled backwards while swimming through the waves on his way back to the beach and his entire catch was taken off his floatline.

We could only guess who or what it was that had done the dirty ...

While a good deal more is known today about great white sharks than before, these magnificent oceanic creatures are still as mysterious as ever. One authoritative source told us that they have 300 teeth in seven rows, bite with a pressure of three tons per square centimetre and can sense your pulse from miles away.

We know all about the teeth, as well as some of their more aggressive traits – the larger great whites, especially – but not a lot else. For instance, when encountering a bunch of them, as Mike Rutzen of Gansbaai's Shark Diving Unlimited has done on numerous occasions, it is impossible to predict which shark in a group will attack and which will not. During a visit to his base at Kleinbaai in the Cape, Mike told Al Venter that whenever he dives with great whites – and he does so a lot – he has to be extremely wary whenever he is in the water with them.

As he states, 'they are unpredictable in the extreme ... they can turn with surprising speed for their size, almost on a sixpence.'

In one of the television documentaries made by Mike for *Discovery Channel*, you see him spending time in the water with a number of great white sharks while they are feeding on a dead whale. The sequence is one of the most astonishing ever filmed, and regarded by one pundit as the 'Everest of shark-diving experiences'.

Though the pack – some of the sharks were four metres or more – was tearing great chunks of blubber off the whale and Mike was right in there next to them, they never spared him a glance. Each one of these creatures was too busy filling its gut to be concerned either with this curious human interloper or the cameraman who filmed the sequence.

And anyway, as he recounted afterwards, 'whale meat probably made for a much better feed than my scrawny torso.'

Not so fortunate was 41-year-old Eric Nerhus, an Australian diver who was swallowed head first by a great white and held in its jaw for two minutes before the shark 'regurgitated' him. That he wasn't killed by the initial bite was miraculous, though he was partially protected by his lead-lined diver's jacket.

At the time this was happening, Nerhus was underwater searching for abalone with his son and other divers off Cape Howe, on Australia's southeast coast. When the shark briefly released its grip and tried to strike again, Nerhus fought back with the heavy metal chisel he used to prise the shellfish off the rocks. He dug the chisel into the shark's eyes and the sensitive area around its snout, which was when the shark released him.

This man simply has to be the luckiest man alive. Of all the creatures on Planet Earth, none is quite so perfectly attuned to killing as the great white shark. In its gaping jaws are rows of serrated, triangular teeth which it uses to rip and tear huge chunks of flesh from its prey, shaking its head from side to side like a terrier with a rat. Nerhus admitted from his hospital bed afterwards that he had never felt fear like that in his life before.

Gletwyn Rubidge of Port Elizabeth is one of the most experienced spearos in the country and stresses that his encounters with sharks are incidental to the underwater experience. Most times he goes out with the intention of bagging something large for the pot. (Photo: Gletwyn Rubidge)

Not everybody is aware that adult great whites sometimes reach six metres or more in length, though there have been reports of this species of shark being caught and measuring in excess of seven metres. Weight can easily be two tons. Though some authorities maintain that their eyesight is poor, one needs to ask, if that were true, how is the shark able to routinely target a victim on the surface while cruising in the depths below – sometimes 20 or 30, even 40 metres deep.

Far from being indiscriminate killing machines, they are actually intelligent, subtle creatures. Mike Rutzen maintains that they are a lot smarter than they are often given credit for. Rather than charge their prey headlong, he reckons, they prefer to stalk it from below. 'Many victims never even see the shark before it hits them, often using the momentum of its upward force through the water to knock a seal or a diver clean out of the water,' he states.

They are also surprisingly cautious. When one of the boats that was viewing great white sharks was overturned by a freak wave off Kleinbaai in the South Cape some years ago, there were six or seven other boats in the vicinity, all chumming for whites and all with one or more sharks waiting to take the bait. About 30 people were hurled into the water and almost all were rescued. Those that died had been trapped in the overturned boat.

Had the whites been the voracious creatures they are rumoured to be, at least some of the sharks would have moved in for the kill. But they never did.

Gletwyn Rubidge takes up the story again:

I have met hundreds of sharks in my 20 years of free-dive spearfishing. Most have simply been passing by and they moved off.

While raggedtooth sharks have always been something of what someone called 'a pestering species' in the Eastern Cape, great whites took a back seat in the early 1990s, but that changed radically in more recent years with a marked increase in encounters with great whites.

Fortunately, the great whites encountered in the Port Elizabeth area are often juvenile and rarely troublesome. I will summarise my views and experiences that relate to contact between humans and great whites.

ENCOUNTER FREQUENCY AND BEHAVIOUR

During the course of the past two decades, I have had 14 direct underwater encounters, or roughly speaking one run-in with whites every two years or so. Only one of these was aggressive, while six swam past and showed only mild interest. Two of these critters circled me and five were cocky enough to relieve me of fish I'd speared – one even biting through a 2.5-millimetre steel cable.

Fellow spearfishermen in Port Elizabeth told me of another 30 encounters, almost all of which involved mild interest, swim-by or theft of speared fish. There were no attempted attacks among any of them. Meantime, juvenile great white shark encounters have increased significantly. In Port Elizabeth two decades ago, we would typically encounter a single great white shark every two or three years. That frequency has risen to two or three *each year* since 2009.

Many of these run-ins involve small great whites in the range of two to 2.5 metres. Fishermen have also reported regular catches of small great whites in Algoa Bay. I encountered a 'small' great white of a little over two metres off Hoby Beach, where the Iron Man swim takes place each year. I prodded that shark in the mid-body with the tip of my speargun and it made off swiftly.

I read of two more being seen in this area and one was caught off the rocks near the aptly named Shark Rock. The fate of the Iron Man Race with its long swim event may change once these smaller sharks have reached full maturity.

FENDING OFF SHARKS

It is axiomatic that like all marine creatures, the eyes and gills of sharks are the most vulnerable. They roll their eyes back when attacking prey or when caught, using a nictating membrane – a translucent third eyelid – that can be drawn across the eye for protection.

I have found that raggedtooth sharks can easily and reliably be fended off by being prodded close behind the eye. I have tested this method over 300 times and it has worked every time. I have twice prodded great whites while they approached and both times the predators sped off and did not return.

Other great whites have backed off when I have shown an element of aggression by moving towards them. The ability to reflect no visible fear is not the norm with these creatures, so when there is something in the water that is unperturbed by their presence, sharks tend to act with caution.

My policy is to prod all the sharks that I encounter in an effort to study their response. That said, I never prod the eye. There is a potential benefit in this approach because it empowers the diver to fend off a shark in such a way that the shark and diver benefit and neither is injured. I also believe that many more tests need to be done to prove the viability on great whites, especially large ones.

The reader can view an example prod on a raggedtooth shark. See the following YouTube clips: http://goo.gl/SJTpU, as well as http://goo.gl/mMX5F.

ORGANISED CAGE DIVING AND CHUMMING

Though cage diving operations with organised chumming often make viable commercial tourist attractions, they do have their pros and cons.

These operations tend to glamorise sharks and also increase awareness, which undoubtedly favours the future of these denizens that are being slaughtered by Chinese interests for their fins to make shark fin soup.

But to my mind, there is a downside. Sharks do respond to stimuli, as do most animals. With time and experience, we have been made aware that sharks have come to associate food (chum) with the sound of outboard motors and waves slapping against the hull of a boat. This association attracts sharks to boats, which are often involved in scuba diving, free-diving and other watersports such as skiing or wakeboarding.

Spearfishermen, for instance, have reported great white sharks surfacing around their boats even before divers got off the boat, especially close to areas where shark baiting is routinely practised. That includes places like Beneke's Klip close to Seal Island in Mossel Bay. Such a stimulus-response could ultimately put people at risk.

The other form of diving where sharks are attracted by employing an artificial bait ball like those used by the operators to attract predators – in this case tiger sharks – is also potentially problematic. Take, for example, the reefs off Scottburgh, immediately to the southwest of Aliwal Shoal on the KwaZulu-Natal South Coast. This was formerly a spot where regular spearfishing competitions were held and there were almost no shark problems whatsoever. Using a speargun, I have 'fished' there for several days in the mid-1990s without spotting a single shark.

In recent years, with shark diving burgeoning and blacktip and tiger sharks the main attractions, the situation has changed. On recent excursions we have found that spearing fish has become almost impossible because sharks – mostly blacktips – arrive in large numbers and swiftly steal almost every fish that is speared. It is obvious that the risk of shark attack has increased. Further, rumours abound of incidents where people have been bitten during shark dives, yet these incidents are not publicised because of the effect it might have on business.

After all, national wildlife conservation areas like the Kruger National Park do not permit the dragging about of carcasses to attract lions to be viewed from close quarters, especially not out in the open where others may be subjected to the increased risk of an aroused predator.

What we do know about great white shark attacks is that they are usually fatal, not only because of the size of the sharks, but the sheer force of their bites.

In an article in Britain's *Spectator* magazine, James Delingpole recently listed several attacks in Australia. He tells us of a surfer off the southern Australian coast who survived an attack with minor injuries, while a 15-year-old boy swimming off a remote south-west beach had his leg bitten. Not long afterwards, a scuba diver off the Western Australian city of Perth survived an attack by a great white after fighting it off, first with his speargun and then with his hands.

A 21-year-old woman was not so fortunate in a great white encounter. She died after she was attacked by three sharks while swimming off an island on Australia's northeast coast, having lost both forearms and suffered wounds to the legs and torso.

An interesting set of statistics here comes from the American State of Florida, a region that annually records the most shark attacks in the United States. In the 15 years between 1990 and 2005, there were 341 shark attacks off Florida, details provided by the American-based International Shark Attack File which can easily be accessed on the web, via Google, if necessary.

It says much that during the same period, Australia reported 74 shark attacks, South Africa's total was 72, Brazil's 62 and Hawaii a significant high of 57 for such a tiny area.

CHAPTER TWENTY THREE

JEFF McKAY TALKS SHARKS

Formerly with the Natal Sharks Board, situated in Umhlanga Rocks, KwaZulu-Natal, Jeff now lives in North Queensland, Australia. Still an avid underwater enthusiast, he does most of his diving these days on the Great Barrier Reef and the South Pacific Islands. In the 1980s, Jeff McKay co-organised great white tagging expeditions off the South African coast with Marie Levine of the Shark Research Institute, Princeton, New Jersey, USA. This is his story of an early encounter in the shark cage.

We were in fairly shallow water off Struisbaai at the time, and from the surface everyone on board the boat whooped at the size of a huge dark form that swam into view. 'Shark!' somebody shouted, and indeed it was. Then this great creature disappeared again, which was when somebody asked where it had gone? That was my cue to bail over the side and make for the aluminium cage that we'd tethered earlier to the stern of Traill Witthuhn's boat. There was always the uncontrolled shudder as the cold Cape water entered into places of my wetsuit. This short swim over to the cage was a blur as my focus was just to get in safely. Then I heard a metallic clang as the shackles were pulled tight against the bars of the cage; the boat was riding the swell at anchor.

Over some days prior to this event, I'd become accustomed to this irregular movement I'd experienced inside the metal cage, and I thought to myself, 'You've developed cage legs', a balancing stance to accommodate the erratic jerky movement of the cage against the securing lines underwater. I also recall listening to my regulator working in tandem with my breathing, which was unusual … I was seldom made aware of it unless it was playing up … acute concentration, no doubt.

Just then I caught a glimpse of movement through the chum at the extreme edge of my vision. A trickle of seawater ran down my face as I tried to identify the source. Moments later a huge white object glided past, which was when I realised that I had focused on the shark's large underbelly.

For a moment or two it seemed incomprehensible that we could have had a great white that close to the boat. I looked at its size and head as well as its enormous girth and my mind froze as I tried to contemplate this enormous majestic shape.

By now the shark had turned and was once again within touching distance. It seemed to grow larger as it came into view. I'd never seen such a big shark before. Looking back, I realise now that I simply couldn't get over its massive girth.

Whether as a consequence of fear, or possibly in anticipation of what might happen next, I shuddered involuntarily as I tried to fiddle with my camera. Relax, dammit, I told myself.

Just then the monster banked towards me and I couldn't believe my luck. I would be able to get a few photos after all, and possibly have the crew tag her. But all that still lay ahead if she didn't move on in the meantime. I pressed on the shutter: one down! How many more to go?

Through my camera's 15-millimetre viewfinder, I saw something I hadn't bargained on. The giant predator was racing towards my cage at an astonishing speed. I could do nothing but back away from the near side of the cage, lift my camera, frame it momentarily and shoot. I'm sure I was looking down the shark's gullet, because in that final stage of the rush, its jaws opened.

With a terrific impact the shark slammed into the corner of the cage, causing me to lose my balance for a moment. Undeterred, because this was a rare experience – in those distant days, at least – I desperately continued to concentrate on taking photos.

With my heart racing, I hugged the camera closer towards my chest. 'What's the matter?' – I was acting like a drunk, and looking up from below, I could see why. The shark had shoved its lower jaw through the bars and was shaking the cage, almost like a rattle.

It was like something out of a horror movie and my brain was having difficulty accepting the reality of what was happening. Then, a more threatening sequence of events seemed to take over and move in slow motion when I saw that with the shark's initial impact, the side door of the cage had opened. Moreover, it was swinging in time with the thrashing of the shark.

I – half hanging out of the cage – urged myself to get it shut, pull my leg back into the cage, latch the top of the door and seek

Andrea Marshall

portfolio

safety behind the metal bars. It must have been in record time, but I performed all those tasks and – shark or no shark – started to feel a lot more comfortable. At least I had achieved a measure of control, even though the sound of the shark's jaws grinding on the bars was disconcerting.

Then, quite suddenly, the thrashing stopped, even though its jaws were still locked onto the corner of the cage and, for the first time, I managed to get a good look at its underbelly. It was only then that I observed that the creature had no claspers: this huge predator was a female, and a very aggressive one at that.

For a few moments we both remained motionless. In fact, time seemed to stand still as we viewed each other, the shark observing my every movement. It seemed like a surreal moment of telepathic communication between us.

What an amazing rush it was. I sensed a hint of empathy, almost a bond with this lady of the ocean. Yet there was no missing how those teeth had slammed and cut into the aluminium bars of my cage.

Contemplating how to get the shark to release its grip before the metal damaged her serrated teeth, I slipped an arm through the bars and pressed against the area just in front of the shark's gill slits. Curiously, I found it all astonishingly firm and couldn't help thinking that the skin had a coarse, almost velvet-like feel about it.

The shark would have none of it and continued to grind her jaws on the cage again. She ignored my touch and the tubing squealed as her teeth cut into the bars. Then finally, almost as if she'd made a decision that I was not going to provide her with a meal, the monster opened her jaws and relaxed. For the first time, my new-found acquaintance looked quite feminine, and with that, the shark slid down to my level and I could see her big black, emotionless eye. It was almost half the size of a saucer and rotated slightly in her socket while giving me another once-over as I hung onto her last and only recognition of our brief encounter.

Slowly, very deliberately, this great lady slipped away from the cage and sank into deeper water.

Meantime, other members of the team had managed to plant our tag into the shark, making it the first great white to be tagged in the region. This was done from the surface when this predator first came alongside the boat.

In the late 1980s, while in the company of Marie Levine, an American colleague and internationally-known shark researcher, we carried out a series of research projects on shark attacks in South Africa. I realised early on from information collected, that if we wanted to study and tag the great white shark, the coastal waters of the Cape and its offshore islands would offer us the best prospects.

During this trip we came across two people who were to become closely linked to the project: André Hartman, a diver who, over many years, has had numerous encounters with great whites while spearfishing, some of which has been flighted on the BBC as well as *Discovery Channel*, and Traill Witthuhn, a professional deep-sea angler and skipper with vast experience of great whites. Both men, I was soon to learn, had a genuine appreciation and understanding of sharks.

A word about the shark, *Carcarodon carcharias*, known around the world by various names, such as the blue pointer, white shark – or, more familiarly, the great white shark. In South Africa, it sometimes goes by the name of tommie, *withaai* and a lot else besides.

This dominant shark of all the world's oceans (they are found in the Antarctic as well, where some years ago, a New Zealand research scientist was killed by one) has been regarded a notorious killer ever since man first became an interloper within its domain.

This predator was first given the name of *tommiehaai* by the Afrikaans-speaking locals of the southernmost part of the African continent. That happened after hundreds of British soldiers – known colloquially among their own as 'Tommies', were brutally savaged, almost all fatally, while trying to swim ashore after their troopship, the HMS *Birkenhead*, had been wrecked off Gansbaai in 1852.

Since earlier reports of shark attacks by survivors or witnesses to these horrific events had been widespread, these predators have consistently generated fear and repulsion. Since the beginning of time almost, mankind has regarded sharks as both dangerous and loathsome. Killing them, until fairly recently, was regarded as the only remedy whenever these creatures were encountered.

I grew up as a young surfer in Durban and was constantly reminded that I should be careful when entering the ocean. I was left in no doubt that shark attacks were a common event along the Natal coast, especially during the late 1950s and 1960s. Most of these were on bathers, but then surfing became more popular, followed by the expansion of the sport diving industry, and some of those participants started to appear among shark attack statistics, surfers especially.

Cumulatively, these incidents led to the formation of the Natal Anti-Shark Measures Board, later renamed the KwaZulu-Natal

Wolfgang Leander

Sharks Board or KZNSB. This operation was initiated under the direction of the late Beulah Davis, an intrepid professional who, as the saying goes, 'got the show on the road'.

According to its charter, the prime function of the KZNSB was to protect bathing beaches using offshore nets, known in our lingo as 'shark nets'. This technique is still widespread and is employed in Australia and South Africa. Indeed, it remains the most cost-effective way of protecting popular bathing beaches, unfortunately at the enormous detriment of the shark because over the years, thousands of sharks have been killed in the process.

Essentially, the process is basic. Shark nets catch inshore roaming sharks at night when these predators are unable to see them. Basically, therefore, this is a fishing device delineated by 'using other means'. The average for great white sharks captured by these nets was in the region of about 50 a year, though the tally tended to fluctuate and could sometimes be much more.

There is no question that the great white shark is a spectacular creature, not only to behold in its natural environment, but also as a superb maritime hunter. Graceful, powerful and deceptively swift when the occasion demands, the great white is an unusually handsome creature, with a sleek, rounded head and extremely prominent black eyes. There are very few underwater enthusiasts who, having encountered the great white in its own environment, have not been awed by its ability as it slices through the water.

The great white's upper body may be metallic blue or grey, sometimes varying to almost black in colour. The underbelly is a sharply contrasting white, broken with patches of black at the tips of the large pectoral fins and sometimes randomly scattered elsewhere on the fin. This gives each shark its own distinctive colouration or pattern, which can be recorded and used later for categorisation by those, like Alison Kock and her colleagues, who spend their professional lives studying them.

Females are the larger of the species and can grow to over six metres, but with the Chinese scouring the oceans for sharks and using longlines to plunder the ocean, larger predators are a rare phenomenon these days. These sharks store food in their livers, which will sometimes result in extensive girths, but during lean times they might shrink markedly to a more slender appearance.

The most common locations for attacks on man by great whites are South Africa, South and Western Australia as well as the west coast of North America. Most onslaughts of late have been on surfers and divers, particularly spearfishermen, or in the argot 'spearos'.

It has been established that many shark attacks in southern African waters have been attributed to great whites, with the more populous zambezi shark possibly tipping the balance.

Some years ago I was involved in developing a method of tagging sharks underwater with Geremy Cliff, an acclaimed marine scientist with the KZN Sharks Board, who has been involved in numerous documentary programmes on these creatures. We worked to a specific system, inserting our tags at the base of the first dorsal fin, usually shot underwater from a modified speargun. This method, when used on the raggedtooth shark, causes no distress and more often than not, the shark is not even aware that it has been targeted.

A method previously used was to insert a tag into the shark by using a lengthy prodding pole from the surface, but that could only be achieved after the shark had been hooked, always stressful for the shark. As a consequence, we resorted to the alternative method.

Then came the time when I believed that the same technique could possibly be employed on great white sharks. However, when we had worked with raggedtooth sharks, the tagging operations were done during breath-hold free dives – as in spearfishing – and we rarely used tanks. But great whites were another proposition, and that was when we decided the best option would be to use scuba equipment, if only for the limited protection more underwater manoeuvrability would offer.

Initially, buckets of ox blood were used to attract these sharks to the boat, though this wasn't always successful. It was only discovered later that using any kind of fish offal – the more odorous the better – was much more successful. It was also more immediate: the sharks would arrive from all over within minutes. Nonetheless, whatever we threw into the water would heighten their primeval feeding instincts and could sometimes lead to a feeding frenzy.

At other times, we discovered that it was possible – in calm and unhurried conditions – to sometimes swim alongside a great white, out of the cage – without being attacked.

Over time, from my experience in dealing with these sharks, I developed the basics of a behaviour guide of dealing with sharks while diving. I found that great whites with dark upper bodies were usually more aggressive and agitated. In contrast, those with lighter pigmentation reflected a more relaxed behaviour. I put it all down to different skin surface tensions, probably due to the mood of the shark, which shifted the reflection angle of the denticles of its skin.

Using this information as both a behaviour and mood indicator – and coupled to a range of colour patterns – I would then determine my judgement in approaching these creatures (or possibly not going anywhere near them that day …).

The question most often asked is why should sharks be tagged? Or phrased another way, why subject them to this 'indignity', as one columnist phrased it? The answer is simple: the process is employed largely to observe and record both their growth and their movements through the sea: modern tags can transmit this information to a satellite which would be downloaded at a specific time for study.

As we have seen in the chapter by Alison Kock – the renowned Cape Town marine biologist who has worked with these sharks for years – the great white is being tagged in Cape waters largely to establish whether the species is semi-territorial and also if local 'resident' great whites restrict themselves to a specific area where there is an abundance of fish, seal or dolphin. Additionally, the programme is geared to find out whether these creatures move farther afield, possibly to other oceans and continents. We now know they do, but only a lot more work and research will produce answers that are relevant.

While still working with the KZN Sharks Board, I was aware of little of this at the time. Two great whites captured in shark nets in KwaZulu-Natal produced a quantity of gut contents of seal from the Cape that surprised us all: the nearest seal colonies were more than 1,000 kilometres away and one of the carcasses had originally been tagged on an island in False Bay.

Which raises the question: how much more has been established since? We now know that some great white sharks tagged in the Cape have travelled to the west coast of Australia and Tasmania, which makes for many more questions than answers.

We are also aware that our southern islands are more abundant in great white sharks in the winter and early spring than in the summer months. That is when pregnant Cape fur seals give birth to their pups. Commercial fishermen and old sealers tell stories of white sharks patrolling during this period those islands and their offshore gullies – such as those lying between Dyer and Geyser islands, a few kilometres off Gansbaai.

Larger, mature great white sharks feed primarily on marine mammals, whereas juveniles seem to prefer fish and small sharks. They also scavenge the carcasses of dead mammals. As the white shark grows larger and its teeth broaden at the base, its diet also changes and it then moves on to mammals, such as seals and dolphins.

Mike Rutzen

In the feeding process of all sharks, there is a definite pecking order. This is based strictly on size (and obviously, aggression and ability). It is now common knowledge that the more 'senior' (larger) sharks will attack their smaller cousins on sight should they not give way when feeding. This display becomes apparent when great whites are attracted to floating carrion, such as whale carcases.

Great white sharks are usually solitary creatures and normally feed alone, but there are also records of groups of them coming together to gorge on shoaling fish or mammals.

When hunting around the islands, they swim along the bottom with their pectoral fins spread wide and making swift tail movements. While adequately camouflaged over reefs or in deeper water, an almost luminescent sandy ocean floor will immediately blow their cover. But wherever the hunt takes place, either on the seabed or on the surface, once a victim has been targeted, the white is able to move in at an incredible speed, reckoned by some to be in excess of 30 kilometres an hour in its final dash.

A shark weighing half a ton will strike its prey with enormous force, especially if the intended victim is a large mammal. Spearos who have been hit by large great whites racing up at them from below are usually lifted clean out of the water. Several of our spearo buddies have suffered serious injury as a consequence and there have been a few deaths.

Interestingly, that first deadly lunge is not to kill, but rather to 'mouth' the victim and establish whether it is palatable: in this process, divers' wetsuits have saved an awful lot of lives. Made of neoprene, more divers have had their lives spared by this synthetic substance than can be imagined: simply put, sharks don't like the taste of rubber ...

That said, once a real attack has taken place, it is not unusual for the shark to back off and wait for its prey to weaken. With severe tissue damage and blood loss, it doesn't take long before the animal returns for the *coup de grâce*.

Great white sharks are also known to cruise around seal colonies, occasionally letting their heads protrude out of the water while spotting for activity along the rocky shoreline. They will react swiftly when a group of young seals take to the water, as they often do, probably believing in 'strength in numbers'.

The bottom line is that in this kind of activity, snorkellers and spearfishermen are the most at risk. Scuba divers are also exposed to unnecessary danger when 'hot-tubbing' alongside the boat before or after a dive, or perhaps floundering about on the surface. This is important when diving, especially if there is an established great white or tiger shark presence in the area. You need to be both aware and alert to these threats with sharks and if in any doubt, it is always advisable to ask local fishermen.

In contrast – and to some people, totally out of character – you might have a situation where divers and members of fishing communities have sometimes come face to face with large sharks and there had been no aggressive behaviour. There is a story doing the rounds in the Antipodes where Arnold Pointer, a fisherman from South Australia, released a captured female great white shark from his net.

As a consequence, Pointer now has the problem of being shadowed by this white shark whenever he goes to sea. Interestingly, he has named the predator Cindy: almost like clockwork, the great beast goes straight to him for the customary pat on the belly and neck whenever he stops his boat. No lightweight, Cindy is over five metres in length and has never been shy to chase after a meal.

This 'affair' has been on-going for more than two years, with an article and pictures about the liaison having appeared in the French publication *Le magazine des voyages de peche* (Edition 56).

Speaking personally, I have had two run-ins with great white sharks while spearfishing in southern African waters, both of which could have turned nasty. The first was with a modest, two-metre shark, which left my speared fish on the bottom of the ocean (where I'd hastily abandoned it in order to scramble to the surface, after having been buzzed).

It all happened quite fast and the only thing I could do at the time was to stare at it and determine what I should do next. To my astonishment it broke away and headed to the ocean floor where it promptly devoured my fish.

The other encounter was a bit more of an adrenalin rush. I was swimming over a reef in about nine metres of water when I turned around just in time to see a three-metre shark charging in at me from behind. I was fortunate that the visibility was good, and I had just enough time to instinctively swing my gun around and take aim.

But before I could pull the trigger, the shark veered past and disappeared into the murk.

Eye-to-eye contact in the wild with any threatening shark – as all divers who have had this experience will tell you – is a challenge.

They are also likely to mention that eye contact is the best, and sometimes the only option. There is no question that when approached in a threatening manner by an aggressive shark, a diver's mask presents the creature with a prominent and, in all probability, intimidating eye contact underwater. Indeed, I firmly believe that should I ever be badly attacked by a shark, it would only happen because I never saw the beast coming. In a nutshell then, situational awareness can, and often does, save the day.

At the same time, it is wise never to become complacent. Any diver in the presence of sharks should always display the aggression of both a hunter and a tracker. In short, never assume you are the only one looking for prey!

Another aspect of this anomaly is that great white sharks – in fact, all the sharks in the ocean – have, within the space of a few short years, become a *threatened* species. Asian fishing boats have cut deep swathes into the shark populations of the Indo-Pacific Basin and many other maritime waters as well. Case in point is the Mozambique coast where sharks have been so severely depleted off Africa's east coast that most divers hardly ever see them anymore. And that is the real tragedy!

Al Venter and his partner Caroline Castell spent three weeks diving between Vilanculos and Zavora early 2012: in all that time they saw a single medium-sized predator off Bazaruto, though there were still quite a few whale sharks around.

As a consequence of this enormous 'marine-disaster-waiting-to-happen', the future of the great white shark – in fact, all sharks – need to be seriously considered, and steps have to be taken at an international level to protect them.

It is axiomatic that senseless exploitation – such a killing sharks for their fins to make shark fin soup – simply has to be stopped. The Chinese have raped Africa of almost all of its rhinoceros as well as a huge proportion of its remaining elephant herds for their ivory: they are now doing the same with the ocean's sharks!

Also culpable of this kind of exploitation are the trophy hunters. Unfortunately, only after they have slaughtered a great white shark and experienced the anti-climax will they begin to understand that it is no big deal to catch and kill one of nature's largest and most threatened predators. In truth, the great white shark – fearsome and ruthless, as it so often is in its feeding process – has its very distinct place in the ecosystem.

To my mind, it is far more rewarding to make the effort, get into a cage and observe the shark at arm's length, and photograph one of your most memorable experiences rather than to try to catch or kill one.

One would like to believe that more pressure is being brought to bear and still more legislation enacted to protect marine mammals all over the world, and that the great white shark will be included in these conservation programmes.

Every responsible diver can make a huge contribution should they put their minds to it. The general public could actually start by boycotting any Chinese restaurant that features shark fin soup on its menu ...

SAFETY TIPS WHEN DIVING WITH SHARKS

1. When entering the water for your dive, always survey the waters immediately around and below you.

2. If you have a problem with your equipment on the surface, have a dive buddy do it for you. There is nothing melodramatic about safety: it makes for good common sense.

3. When descending, maintain a 360-degree lookout.

4. During the dive, keep situational awareness and do a regular 360-degree scan.

5. When surfacing or decompressing, it is extremely important to keep a vigilant lookout.

6. After you arrive at the surface, always keep a wary eye out for anything happening below: continue doing so until you leave the water. Try to get onto the boat as soon as possible after a dive.

CHAPTER TWENTY FOUR

FOUR YOUNG MEN IN MOZAMBIQUE WITH VISION

Sharks are being plundered at an alarming rate off the Mozambique coast by Chinese longline fishermen: tragically, they have all but stripped the Indo-Pacific Basin of its sharks. But that does not mean that divers have the 'all clear' when in open seas, as Morgan O'Kennedy discovered on a recent spearfishing trip off the Bazaruto Archipelago.

As Morgan recalls, he and Michael Klue – one of his business partners involved in their Big Blue dive operation at Vilanculos – decided they needed a break. The idea was to fill the tanks in one of the company boats with fuel and drag along two female acquaintances for a day of spearfishing.

With a bunch of fishing rods, free-diving gear and refreshments stowed, they headed out for deeper waters and the San Sebastian area. As Morgan recalls, it was the start of the summer and a particularly flat day; the four of them were soon treated to an array of sightings and encounters with marine life. The dolphins they saw that morning were spectacular, as happens so often along the length of the Bazaruto Archipelago: they had pod after pod playing in their bow wave. Morgan takes up the story:

With the sun only just up and lying too low on the horizon to really see a lot in the water, we started off with the rods and after a couple of hours of playing with the lines, we managed to hook a nice king mackerel for dinner: we released another two of the species as well as a wahoo. Our female companions were soon eager to try something else, like looking for whale sharks with which they could frolic.

We didn't have to go very far and soon found two of these magnificent creatures in reasonably shallow waters. We went

Vilanculos Harbour – A delightful seaside town in Central Mozambique, Vilanculos is the gateway to Bazaruto and reflects much that local custom along this stretch of coast is known for, including Arab dhows. (Photo: Author)

overboard and were able to free-dive with them for about half an hour in the kind of crystal-clear water often encountered in this region. It was, as they say, a truly magnificent day. By then Michael and I were considering shooting some big trophy fish.

While still swimming with the whale sharks, we were again joined by a pod of bottlenose dolphins – my first snorkelling experience with these mammals. Brunch followed soon afterwards and we decided to move on; the idea was to head to deeper reefs and take out our spearguns.

It was not yet noon when we went in with the guns. Visibility was over 30 metres and there was no missing the Pinnacles and surrounding reefs that stretched away before us like a fairyland apparition. Early on, we'd decided that we would only shoot a fish if it was trophy quality – something we could take home with us. In so doing, we reckoned we'd keep the groupers in their holes, the sharks at bay and the game fish relaxed.

Michael shot a mackerel of more than 15 kilograms in almost record time and got it safely back on the boat. Some pretty big wahoo followed, but nothing too exciting. After about eight or ten drift dives over the first pinnacle, we both mentioned the absence of sharks, which, anyone who has dived there, will say is a bit unusual for Bazaruto. By now the girls were getting a bit restless, so we hauled across the bay to a shallower reef so they could also get in some free-diving time. After about an hour of checking out turtles and plenty of reef fish, we were again ready to head deeper; the prospect of trophy fish was tempting.

We did a couple of drift dives and saw little that excited us and it was then that we decided to shoot some fish for our boatman, the gillie. The fellow driving our boat and the workers back home deserved their share of the takings, which is something we always do when we go out.

I was first in the water and was quickly greeted by the spectacle of hundreds of yellowspot kingfish. There was also bludger and big-eye kingfish, all doing their shoaling dance just ahead of the boat. I took a deep breath and swam down towards the middle of the shoal, selected the largest yellowspot within my range and pulled the trigger.

What often happens in very clear water is that you sometimes tend to misjudge distance. This is what happened and it also meant that my spear struck the fish a little too far back on its torso. It was a nice specimen of almost 10 kilograms and still very much alive … in no time at all the fish was able to take my gun and most of the line connected to my buoy straight to the bottom.

At that stage I wasn't too worried. We hadn't seen any sharks and I thought we were safe, but then nature sometimes provides its own set of surprises. Almost as soon as the thought crossed my mind, a

The beautiful peninsula on which is perched Nyati Beach Lodge – the lagoon and estuary lie on one side and the Indian Ocean stretches out expansively on the other. (Photo: Janneman Conradie)

small zambezi shark of about 60 kilograms raced up to my 'catch' and within seconds it had gobbled up half of it. That done, the shark decided it still hadn't had enough and returned for what was left of the tail.

If you've ever dived these waters, you accept that this sort of thing happens and that there is no need to get too excited about it. What is essential is that you remain calm. As they say in the game, you win some you lose some … that's the way it was just then …

I pulled my speargun back towards the surface once more and prepared for the next dive. But I'd barely recovered five or six metres of line when I suddenly saw the biggest zambezi shark of my life … it was easily three metres plus and was swimming right below me at about 20 metres. It was obvious that the predator was heading towards the few remaining pieces of fish, scales and guts that the smaller shark had left behind.

The newcomer seemed relaxed and, I thought, probably only coming to see what the commotion was about, and that was when I suppose it got a whiff of the kill. Quite unexpectedly, I witnessed the shark's body language and posture change, almost as if it had been given a signal. I've dived enough to be aware that there was trouble coming our way.

Having scoffed the remains of the kill and after doing two more small circles, this very large zambezi shark turned towards where I was hovering on the surface. It didn't hesitate a moment ... just kept coming straight towards me ...

The last thing I remember before it hit me in the lower regions was the splurge of white on its underbelly. The next moment I was lifted clear out of the water by the shark and hurled through the air. I landed back in the water with a splash, my mask at the back of my head. One of my fins had been dislodged and my body was entangled in rope. What a pickle, was all I could think just then.

I didn't waste any time getting my mask back on straight, which was when I saw the shark again, this time on the surface about five metres away. It moved slightly in the other direction and then came in at me again.

By now the gillie was very much aware of what was going on. He'd seen me do a double flip after the shark's blunt nose had struck my leg and keenly followed the action as I swam strongly in his direction. The shark was there as well, following closely in my wake. As he recalled afterwards, he thought it was going to get me the second time round.

I reached the boat soon enough and, as I pulled myself on board, it felt as if I'd just been rugby-tackled by a heavyweight at speed. I couldn't feel either my right buttocks or my upper leg, which told me one of two things: dead leg or no leg at all.

Some shark attack victims often say afterwards that initially they didn't actually feel any pain after being hit. I was aware, too, that quite often the damage is not visible at first, because wetsuits tend to hold things together, wounds especially. My only thought just then was that the shark had got my leg and I was waiting for the pain and blood. But nothing happened.

In my desperate bid to check the damage, I did a hasty strip and ended up totally naked in front of the two lovely girls. Then it became clear: I wasn't really hurt and there was no blood ...

After that powerful rush of shock, I suddenly felt my knees buckle: I couldn't stand anymore, which was when I lay down on the deck and stayed prone till we got home.

It was interesting, when trying to analyse the situation a day or two later, that though I'd gone into the water with my usual dive knife strapped to my leg, I wasn't actually able to defend myself.

Everything happened too swiftly, but more significantly, those larger sharks are fast, agile and possess astonishing power.

The creature was a big bugger and it could easily have killed me had it intended to do so. But it didn't. Which raises the question – why not?

Perhaps the zambie was territorial and wanted me out of its area. Or, more likely, it was 'mouthing' and didn't press home its advantage after it had tasted neoprene. We will never know.

What is also true is that since then, I haven't again speared fish in deep water ...

Morgan O'Kennedy's Mozambique adventures started a few years before, when he joined forces with three of his friends from Stellenbosch University to launch what is probably one of the most successful dive and coastal safari operations in the former Portuguese colony. These were fellow students Janneman Conradie, Michael Klue and Pieter Scholtz, each one of them like-minded outdoorsmen and diving enthusiasts.

Janneman Conradie took this shot from his microlight of a whale having just given birth to a calf within the confines of Bazaruto. The calf is resting on its mother's back.

Their company, Big Blue, is based on the water's edge in the central town of Vilanculos and runs dive as well as conservation-oriented operations up and down the coast. Their activities take them as far north as Pemba near the Tanzanian frontier, quite often south towards Zavora, where Jon Wright and his equally remarkable Moz Dive operation sometimes joins forces with them. In April 2012 O'Kennedy spent time on a transoceanic safari to Bassas da India.

It hasn't taken long for Big Blue to emerge as an unusually visionary dive group, both in its reserves of energy and enthusiasm. Each of the four young men is responsible for a specific aspect of its organisation. O'Kennedy runs the lodge, boats and vehicles at Vilanculos, while his immediate partner, Janneman Conradie, concentrates on conservation and keeping poachers at bay. He has also taken it upon himself to try to protect Mozambique's dwindling dugong population, which is now reckoned to number less than 100. This is no simple task in a Mozambique that is still struggling with enormous 'growing pains' following the end of the RENAMO civil war.

Conradie added his additional two bits' worth by bringing into the equation several Micro Aviation Bantams – high-wing, two-seater microlight aircraft built in both New Zealand and South Africa by Micro Aviation. The role of these machines has made Big Blue unique along the shores of the Indian Ocean. They not only assist Conradie

One of the microlights routinely flown by Janneman Conradie from his Vilanculos base: these are two-seaters and visitors can fly with him for a fee. (Photo: Author)

in countering poaching but are also used for spotting fish shoals, dugongs and whale sharks, as well as helping Dr Andrea Marshall gauge the presence of giant mantas, her NGO speciality along these shores for the past decade. But more later about this 'Manta Queen' – as she was dubbed in the recent one-hour documentary produced by the BBC about the activities of this remarkable young American scientist.

Interestingly, Big Blue is due to take delivery of the first amphibious 'float undercarriage' FIB (flying inflatable boat, or 'Flying Duck') version of the miniature aircraft in southern Africa: it has a wingspan of almost 10 metres and a fully loaded take-off weight of just on half a ton. What is certain is that Big Blue is likely to revolutionise all maritime sports in these waters and their operations can be accessed on the web at www.flyinginflatableboats.com.

To date, Janneman has flown his little machines thousands of kilometres up and down the Mozambique coast, but not without incident. He was twice forced down by mechanical failure, once in the ocean, several hundred metres from shore where, in pitch-black darkness, he had to salvage his broken plane in the surf totally on his own by swimming back and forth to the shore. Undeterred, it was only months before he had another of these machines operational again and was running anti-netting patrols in and around the Bazaruto Archipelago, where gill nets are illegal.

He also spends a lot of time checking on the activity (and illegal plunder) of Mozambique's dugongs, which are now seriously threatened with extinction by poachers, whom he regards as both mindless and ruthless, and also dangerous. He has had more than one threat made against his life.

The third member of the quartet is 26-year-old Michael Klue. Having grown up on the family wine farm, he spent a few years studying the liquor trade, with the intention of taking over with his brother after his father had retired.

'But it wasn't for me,' he declares with a smile. 'When you've spent time under the deep blue sea like we have – we're all divers and spearos – then you never want to do anything else ... and now we've got microlights and that gives us another dimension altogether.'

Today Klue's ancillary role is as general manager of the ultra-luxury Nyati Beach Lodge that lies perched on a hilltop overlooking the ocean about 60 minutes by boat south of Vilanculos. The last word in splendiferous tourist comfort, the 26-bed Nyati Beach Lodge is regarded as one of the best resort hotels in southern Africa.

Pieter Scholtz makes up the fourth member of the team and his role for the time being, is as a general factotum to the rest of the gang, but that is only when he's not diving professionally in the oil industry off Africa's west coast.

Asked what brought them all together in a country that none of them had visited before working there and that didn't even use English as a first language – Mozambique is Portuguese speaking, with Swahili a close second in the north – the four young idealists are explicit.

Said Klue: 'While at varsity, we spent many weekends spearfishing off the Cape coast in places like Arniston, Struisbaai and elsewhere, so when this Vilanculos opportunity came along, we didn't even have to think about it. In fact, we grabbed the opportunity, though it was tough to start with.

'We kind of sidled in sideways and had a bit of luck and unexpected support from a few individuals.'

'Big Blue's lodge at Vilanculos was originally bought from an owner who had allowed everything to go to seed after being struck by one of Mozambique's periodic cyclones in 2007. The owner ended up losing interest and we took over – the place was a total mess.

'So we had to step in and renovate everything from the ground up. That took time and effort.' In the process, he said, they had to make contact with the government, as well as local administrators and face up to demands required of them in what was a totally new venture in a Third World country.

'We actually had to show good intent, so that they, in turn, would make things happen,' said Michael Klue. He intimated that it was a difficult path to negotiate because there had been so many fly-by-night operators who had come in after the civil war with all sorts of promises, which they couldn't deliver. Some of them are still around: one enterprising old 'yachtie' entered the country six years before on a 30-day tourist visa and was only given his marching orders while we were visiting the place. The authorities were actually quite good about it: they could have thrown him into jail and seized his boat, but they did neither. Instead, they gave him a week to leave the country.

'And then, there was the sizeable risk factor. We went in cold, into a foreign country that had recently emerged from a civil war ... it involved investing a lot of money that we had to borrow and then repay ... it took a huge amount of effort and more than a modicum of trust, goodwill, commitment and, as O'Kennedy admits, some solid dollops of luck.

'My first job,' says Michael Klue, 'was to handle all the administration – the paperwork and the kind of bureaucracy that characterises a developing country like this ... and that's a difficult call, even today. But we managed, and in 2009 I quickly learned how to get a company on its feet. Moreover, we did so in spite of a significant lack of support from local banks ...'

Part of this illustrious quartet's job was to knock the big adjacent hotel complex Aguia Negra back into shape after it had been damaged by high winds. 'We fixed it up good, even had to re-thatch many of the bungalows because we knew that we would get a lot of business from there, which, of course, we have.'

Along the way, both he and Janneman will tell you, they did

quite a lot of consultancy work, helping South Africans and others to invest, hopefully wisely, in an environment that was fraught with problems, including a slew of fast-talking con artists. He credits many of the government people with whom he came into contact for much of the goodwill: 'that helped a lot,' he says, adding that they couldn't have done without it, 'us all being young and pretty inexperienced.'

Janneman: 'We selected our Mozambique partners carefully and, as trust developed, we all started to score. But it couldn't have been done alone ... nothing would have happened without their help ... our Portuguese is simply not good enough ... the labyrinth of red tape and volumes of documents we had to complete, legal matters, attestations and the rest.' As he will admit today, he is astonished they stayed the distance.

In the end, they admit, it was worth the effort and many sleepless nights: Big Blue as a dive, fishing, boat charter, microlight, conservation and exploratory venture is firmly on the map.

How spearo Michael Klue, at the relatively young age of 25 got into the enviable position of being appointed general manager of Nyati Beach Lodge, is a story in itself. As he says, things just happened ...

The place has four specific seasons, one each for sailfish, another for marlin, October to April for whale sharks and giant manta rays, and six solid months for turtle breeding. It also boasts being the most isolated of the lodges on the archipelago, needing more than an hour

Wolfgang Leander

by fast boat to reach the peninsula which it tops, the lagoon-cum-estuary on one side and the rolling Indian Ocean on the other.

Diving is obviously a significant attraction for anybody going there, with additional easy access to the Vilanculos Coastal Wildlife Sanctuary. The place has already attracted attention, with Tom Daily, the noted American photo journalist, describing adjacent reefs in the magazine *Discover Diver* as 'without a doubt, the most untouched and pristine reefs in the world'.

Klue recalls how he and the others discovered the site before it had been properly developed:

'We'd go out there after a day's work and find a place to moor the boat. In minutes we'd go over the top of the sandspit and get into the water, which was almost totally pristine and alive with all sorts of marine life. It didn't take us long to find the best reefs, which meant that we would sometimes go across the bay for a couple of days or more and make camp.

'But we knew, too, that we were on private property, so we kind of camped around the corner from the hilltop lodge where the owner Conrad Pretorius had his base when he wasn't at home in Nelspruit. It got so good that we started taking people with us to share these delights.

'Then, one day we bumped into Conrad and he was quite friendly, intimating that he had been aware all the time of what we'd been

Dr Andrea Marshall's headquarters on the grounds of the Casa Barry Lodge in Tofo.
(Photo: Author)

doing and asked us why we were so secretive? He told us that we only had to ask and he'd have no objection to us being on his land, especially since he shared the same interests as us.

'So, one thing led to another, he made me an offer I couldn't refuse and today I'm running his show. It might sound glamorous, but it is hard work, especially when you're competing against some of the really big international venues that regularly appear in the British media like Vamizi in northern Mozambique and the beautiful Bazaruto Lodge, just around the proverbial corner.

One organisation with whom Janneman Conradie, in particular, has good contact is Dr Andrea Marshall's Marine Megafauna Foundation (MMF).

Originally founded as an association and headquartered at Tofo – down the Mozambique coast from where the four South African idealists have their base – the body was originally geared to research, protect and conserve the large populations of marine life found along that stretch of East African coastline. These included sharks, rays and marine mammals such as turtles and dugongs, all key components of a complex array of international marine ecosystems. The trouble is that since most are long-lived, they have low reproductive rates, and their populations are the first to be depleted by human interference such as has been happening with illegal Asian fishing off Mozambique's shores.

A remarkable young woman, Dr Marshall was 22 when she first arrived in the former Portuguese territory little more than a decade ago totally on her own. Along the way she encountered much scepticism, though she was sometimes helped by ordinary folk who could appreciate the long-term purpose of what she was trying to do. Obviously her passionate commitment to the eastern African coast helped. It also fashioned her goals as a conservation biologist. In time, she was joined by Richard, her soul mate in this extremely diverse activity: he is heavily into construction along the coast.

Having vowed to dedicate her life to the preservation and management of the manta ray population in southern Mozambique, Dr Marshall continues to campaign for their protection and to use her scientific background to formulate plans for their management. Her website can be accessed at www.marinemegafauna.org.

In this respect, Janneman Conradie and his microlights are a valuable adjunct to aspects of what she does: he is in the air almost every day, most times offshore and is able to pinpoint migrating shoals and also indicate numbers, not only of mantas but other large oceanic creatures. One of his most striking photos is of a female whale that had just given birth, its almost albino calf perched comfortably on her back in fairly shallow water.

Educated in the United States and Australia, Dr Marshall was the first person in the world to complete a PhD on manta rays. After finishing her thesis in 2008, she stayed on in Mozambique to spearhead conservation efforts of this species along this still-remote coastline, fully aware that there were forces at work trying to deplete their numbers.

Dr Marshall learned to dive at a very young age, having qualified at the relatively tender age of twelve. Thousands of dives later, she spends time underwater both for pleasure and her profession as one of the leading marine field biologists in southern Africa. She delights in being regularly in touch with her other female scientific and photographic friends living and working in southern Africa, including Alison Kock (who works with great white sharks in the Cape) as well as that dynamic 'Shark Warrior', Lesley Rochat, another of the prominent conservationists of her epoch who wrote the Introduction to this book.

Indeed, to dive with her is a delight, especially when she is in what she likes to refer to as her 'tagging mode.' Caroline Castell spent a week with her at her Tofo headquarters and was present at one such a fixture. She describes the operation as follows:

'There were three others in the water with Andrea when she spotted a giant manta emerge from out of the blue within five or six metres of our boat. Janneman had followed Andrea into the water and then I went in with Dan, one of her resident scientists.

'Initially the manta swam off, but Andrea was resolute and headed after it while the rest of us hung back so she could complete the task. She finally caught up with it, approached from the rear and above, and in a single swift movement fixed the tag to the animal in the thick layer of skin where the manta's wing is attached.

'As Andrea explained afterwards, that was where the barb (to which the tag is attached) is required to remain intact, hopefully for months. Eventually the tag will come free, float to the surface and release all its information to a satellite.'

It is worth mentioning that while Caroline and Janneman were at Tofo, Dr Marshall received input that had been released in 2009. She had originally targeted it at Tofo, after which the creature headed south, towards Durban. Then it turned north again, the entire routine having taken place over several weeks, after which the tag was dislodged. The most astonishing revelation that emerged from its accumulated data was that the manta was recorded at one stage as having descended more than 3,000 feet into the depths of the ocean. At the time this made it the deepest dive ever recorded by a tag on any of the oceanic creatures, though that has since been eclipsed by a whale shark.

One of the anomalies to emerge from Dr Marshall's work in the area in which she is active, is that about 70 per cent of all the giant mantas encountered, have had chunks bitten out of them by sharks, which, she admits, is unusual. She has also lately had the

One of the clandestine shark finning operations hidden in the bush off a southern beach in Mozambique. You could walk within metres of it and not be aware of what is going on there.
(Photo: Author)

Cutting the fins off sharks on a Mozambique beach: tens of thousands of sharks are slaughtered each year for no other purpose than to remove their fins, which the Chinese use to make soup. The rest of the carcass is discarded, often thrown back into the ocean while the creatures are still alive. (Photo: Mike Rutzen)

misfortune of having to remove spears from some mantas that have been targeted by Mozambique spearos who have been taught the sport but sadly, none of its ethics – this is a sad indictment of an activity gone berserk among people who use the guns to put meals on the table for their families and end up stripping huge sections of Mozambique's most productive reefs.

As with other conservationists active in the Indo-Pacific Basin, Dr Andrea Marshall is appalled at the activities of Asian fishermen who make an industry of killing sharks for their fins. These goings-on are endemic in Mozambique, with well-organised and lucratively sponsored gangs using large groups of impecunious locals to achieve their aims.

One area that has been active for a while is along an isolated and quite magnificent stretch of beach at the tiny coastal village of Mahangate, almost midway between Tofo and Vilanculos.

At Janneman's behest, Caroline Castell spent a while there as well, and was able to observe at first-hand how this despicable system operated. Nor was this the first time that these operators had been checked: earlier efforts at stopping this activity only resulted in the shark cullers moving into a more remote area a little further up the coast.

As Caroline says, it was a most disturbing experience. 'While those involved spoke quite freely about killing large numbers of sharks, many of the rotting carcases were simply abandoned on the beach, which reeked of destruction.

'Yet,' she recalls, 'in one corner of the bay, there stood a brand-new solar-powered freezer, into which the fins were deposited each time the crews returned to shore.'

There was a set routine, she discovered. Once a week somebody would load up a truck with the booty and drive to Massinga, the nearest town on Mozambique's north-south Highway E1. A functionary working for Chinese interests – not always Chinese, but sometimes Asian or somebody important in the region – would take delivery and hand over the cash.

'It was like clockwork, and you don't get involved because people that have become too intrusive in the past tend to have accidents,' she declared.

What was disturbing, Ms Castell added, was that this same process is taking place dozens of times a day, year-in and year-out, along the entire length of Mozambique's 3,000 kilometre coastline. Obviously, an enormous number of sharks is being slaughtered.

That is also the reason why, during our own three-week diving venture in Mozambique in April 2012, the author and Caroline Castell saw only a single live predator underwater during the entire period: it was a juvenile shark of about two metres on Twelve Mile Reef, northeast of Bazaruto Lodge ...

ACKNOWLEDGEMENTS

There have been many good souls who have helped in the preparation of this work, which I've been working on for several years. Tackling a subject as demanding as diving with sharks has its own little quirks, not least that there is a fair section of the population that thinks you're a bit loopy getting into the water with them in the first place, especially without the protection of a cage. Truth is, I have only been in a shark cage once, and that was with Jeff McKay and Marie Levine off the South Cape in the 1980s. Though we chummed for a week, there were no sharks …

This book is special in many respects. It would have been impossible had I not been helped by people like my old dive oppos Peter Sachs, André Hartman, John Visser and Walter Bernardis. Geremy Cliff of the Sharks Board, who I have known for decades, has two insightful chapters in this work. Others involved who offered support were Jerry Buirski, Sijmon de Waal, Tommy Botha – and his dive partners Gavin Clackworthy and Steve Valentine – Darrell Hattingh, Sean Botha, British-based Morgan Riley (who tells us all about whale sharks), the indomitable Len Jones – in his mid-70s and still diving – Andy Cobb, Barry Skinstad, Gletwyn Rubidge, former National Parks Head Ranger Peet Joubert, Mark Webster and Colin Ogden, who played a significant role in my previous underwater book *Dive South Africa*. Debbie Smith completed a lovely section on the 'Sardine Run', as well as Hermanus's 'Terrible Twins', Brian MacFarlane and his old chum 'Willie-wil-nie-werk-nie' van Rensburg. Willie survived a serious attack by a great white and is lucky to have survived: his story has been included. Andrew Woodburn and Peter Pinnock, both of Durban, featured prominently in the last dive book and their underlying presence is manifest. I thank you both.

Over almost half a century of diving, many other personages come to mind, including some of my old pals at the Atlantic Underwater Club in Cape Town, including Tubby Gericke, and Brian Clark, both now departed. It was only after I'd joined the AUC that I first met the man who, for many years, acted as a beacon in my underwater efforts. That was Albert Falco, who I first met when Jacques-Yves Cousteau's *Calypso* called at the Cape. He became solid support for the old man and, with Falco's death earlier this year, two of the diving greats have left us. Ron and Val Taylor were also guests at the club when, with legendary Peter Gimbel and Stan Waterman, they explored the possibility of making *Blue Water White Death* in South African waters. More than 40 years later, the film remains a classic.

Very special thanks must go to several individuals who played a significant role in bringing this volume to fruition. That includes the Ramsgate-based couple Roland and Beulah Mauz who have turned shark diving completely around on Protea Banks. I regard that reef as arguably the best location for shark diving on the continent of Africa. Then comes the enterprising Vilanculos-based quartet who have managed to create for themselves and others a series of remarkable opportunities at their Big Blue dive facility in Mozambique: Janneman Conradie, Morgan O'Kennedy, Michael Klue and Pieter Scholtz, all formerly of the University of Stellenbosch.

Lesley Rochat deserves a section of her own, not only for the beautiful portfolio of photographs she came up with, but also for having the courage to tackle the Sharks Board head-on for what she believes, right or wrong, they are doing to sharks. Like her good friend, American-born Dr Andrea Marshall – who has spent a third of her young life trying to protect manta rays (and also gave us a lovely selection of underwater photos) – she tends to bang her head an awful lot, but neither of these enterprising young women is deterred. So too with Fiona Ayerst, a leader in her field with her own portfolio of photographs which, like the others, is magnificent. Her partner, Ryan Johnson, came up trumps with the chapter on the Mossel Bay shark research facility which he founded.

The fourth photo portfolio comes from a remarkable young man who, for a long time, made his money filming great white sharks underwater, *without* the protection of a cage. Morné Hardenberg is married to Alison Kock who, with her monumental great white shark tagging programme, is another leader in her field: obviously the two marvellously complement each other. Alison's chapter shows what one individual – with dollops of grit and guts – can achieve when she puts her mind to it. She acknowledges it would probably not have got this far without Morné's support: blessings on both their heads.

Other notables include Graham Powell, Mike Rutzen, Boetie Scheun, who relocated to Hermanus from Namibia, Alex Pappayanni, Kevin Graham, last heard, diving at Vamizi, my old buddy Shane Breedt who creates his beautifully crafted Freedivers spearguns and is himself a spearo of note.

At the end of the day, this book would not have emerged without the help, support, good wishes and beautiful shark photographs of my dear old friend Wolfgang 'Wolfie' Leander, who lives on the Pacific, almost at the other end of the globe. He is a regular visitor to southern Africa and, as with efforts at shark conservation in other regions (the Bahamas), Wolfie can be acerbic when it comes to trying to protect his beloved tiger sharks. You're a brave man, my friend, and a pox on the homes of those who ignore your sensible entreaties.

Felix Leander, Wolfie's son who lives in Florida in the United States, has also provided some of the images used between these covers. My grateful thanks to both of these outstanding undersea lensmen.

On the production side, Danél Hanekom, with whom I worked on *War Stories by Al J. Venter and Friends* – my most successful South African book to date – came to the fore again as editor and helped put this volume together. Carmen Hansen-Kruger had her hands full proofing it: I am indebted to you both.

Finally, there is my old compadre Peter Younghusband, who originally taught me how to write the kind of tight copy that Fleet Street once liked. As the *Newsweek* and London *Daily Mail* man in Africa for decades, he had good experience to draw on. He's still around, with a regular column in the local rag, and as always, a marvellous raconteur. Whenever I go to the South Cape, either to head out on the shark viewing boats or dive on HMS *Birkenhead* (which I did a lot in the old days) I always stay at his B&B at Stanford, which I regard as the best value for money in southern Africa. A patron of the arts, Peter's gallery is across the road and the beer at his 'local', a short stroll down the road, is always cold. Try it – you won't be disappointed: it is only a ten-minute drive from the shark boats.

As usual, my very good friend and master graphics artist Bruce Gonneau put this book together, and like all the other titles he has handled on my behalf, he has done an extremely professional job of it. Thanks Brucie – and thanks too, to Jenny, for putting up with all his nonsense when he is getting down to it. Here's to the next title we work on together ...

Al J. Venter
Surrey Hills,
July 2012

In the same series published by Protea Book House:

A parachute battalion strike in Angola; clandestine security operations in Botswana; Portugal's ill-fated colonial wars in Africa that lasted twice as long as America's conflicts in Vietnam; hunting big game in an Africa at war and an American hostage with Hizbollah, who was later executed …

All these and more, make up Al Venter's latest collection of war stories, compiled with the help of some of his friends, many of them recounting personal adventures that have never appeared in print before. Like his other books, all make for compelling reading.

War Stories is a significant and timely collection of military events. The book spans Africa, the Middle East and Asia and is illustrated by more than 100 photographs.

War Stories: 978-1-86919-410-9

By the same author:

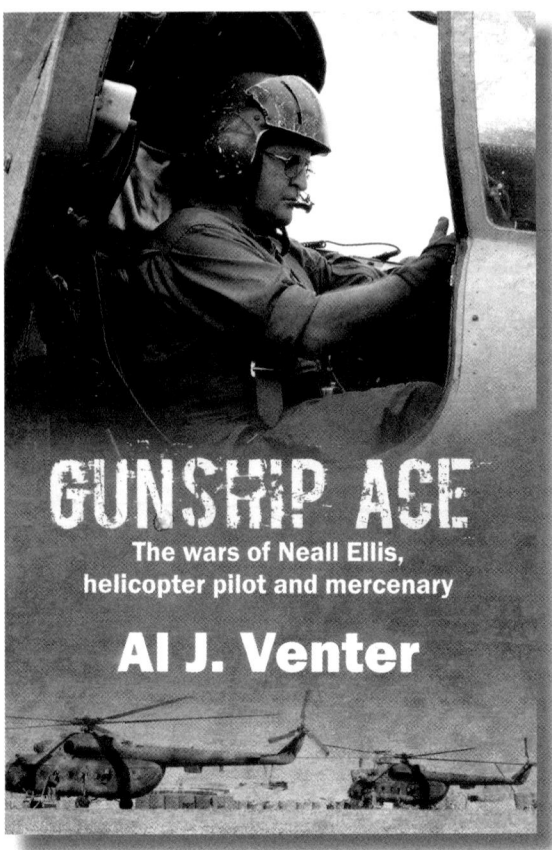

'Neal Ellis is the best-known mercenary combat aviator alive.' This is according to Al J. Venter, who accompanied the former South African Air Force pilot on some of his missions.

Apart from flying Alouette helicopter gunships in Angola, Neal Ellis has fought in the Balkan War, tried to resuscitate Mobutu's ailing air force during his final days of ruling the Congo, flew Mi-8s for Executive Outcomes and finally an Mi-8 fondly dubbed 'Bokkie' in Sierra Leone. Twice, he single-handedly turned the enemy back from the gates of Freetown, preventing the rebels from overrunning the capital – a dangerous task for which many owe him their lives, and for which others placed a reward of millions on his head, dead or alive. Numerous times he flew SAS personnel on reconnaissance missions into the interior of the diamond-rich country, for the simple reason that no other pilot knew the country – or the enemy – better than he did.

This book describes the full career of a legendary aerial warrior, from the bush and jungles of Africa to the forests of the Balkans and the merciless mountains of today's Afghanistan. Along the way the reader encounters examples of incredible heroism for hire, as well as a multi-ethnic array of enemies ranging from ideological to cold-blooded to pure evil.

Gunship Ace: 978-1-86919-703-2